PRACTICAL GEOSTATISTICS

PRACTICAL GEOSTATISTICS

ISOBEL CLARK

B.Sc., M.Sc., D.I.C., Ph.D., C.Eng.

Lecturer, Department of Mineral Resources Engineering,
Royal School of Mines,, Imperial College of Science and Technology,
London, UK

ELSEVIER APPLIED SCIENCE PUBLISHERS
LONDON and NEW YORK

ELSEVIER APPLIED SCIENCE PUBLISHERS LTD
Ripple Road, Barking, Essex, England

Sole Distributor in the USA and Canada
ELSEVIER SCIENCE PUBLISHING CO., INC.
52 Vanderbilt Avenue, New York, NY 10017, USA

British Library Cataloguing in Publication Data

Clark, Isobel
Practical geostatistics.
1. Geology—Statistical methods
I. Title
519.5′02′4553 QE33.2.M3

ISBN 0-85334-843-X

WITH 22 TABLES AND 75 ILLUSTRATIONS

First Edition © 1979
Reprinted 1982
Reprinted 1984

Printed in Great Britain by Galliard (Printers) Ltd, Great Yarmouth

‘*What is obscure is seldom clear.*’

Edwin Morgan

Preface

This book is aimed at postgraduates, undergraduates and workers in industry who require an introduction to geostatistics. It is based on seven years of courses to undergraduates, M.Sc. students and short courses to industry, and reflects the problems which have been encountered in presenting this material to mining engineers and geologists over a wide age range, and with an equally wide range of numerical ability. The book would provide the foundation of a course of about 20 to 30 hours, or of a five-day short course.

The level of mathematical and statistical ability required is fairly rudimentary; it is sufficient to be able to cope with concepts like mean, variance, standard error of the mean, normal and log-normal distributions, and to have some notion of the background to solving sets of simultaneous equations.

As an introduction to a subject which is commonly presented as rather complex, the book will familiarise the reader with the concepts and techniques of geostatistics, providing the necessary foundation to enable him or her to evaluate basic idealised examples. It also gives an indication of how to employ the techniques in more complex and realistic situations.

Geostatistics is used throughout this book in its European sense of the 'Theory of Regionalised Variables', developed by Georges Matheron and co-workers at the Centre du Morphologie Mathématique at Fontainebleau. Although most of the examples are drawn from mining, this reflects the distribution of practitioners rather than the potential of the techniques. Practically any problem which involves the distribution of some variable in one dimension (e.g. time series), two dimensions (e.g. rainfall), or three dimensions (e.g. disseminated ore deposits) can be solved using such a technique.

The units of measurement used in the book also reflect the state of the mining industry. No attempt has been made to standardise to SI units. The examples are real examples, and it seems absurd to turn, for example, 5-ft cores into 1·52-m cores. The one case where this *seems* to have been done is an actual example where the mine involved had adopted SI units, but continued to use its pre-metric 5-ft core boxes.

In the presentation of the material I have tried to show how the basic ideas may be developed intuitively, and I have tended to avoid supporting the ideas with a rigorous mathematical derivation, since there are numerous existing publications which use this latter approach almost exclusively. While many computational difficulties can be eased by use of computer programs, such assistance should not be needed within the scope of this text. None of the examples is too unwieldy for pencil and paper, far less for a calculator. Where formulae are too complex (or tedious) to calculate by hand, tables have been provided. One example has been included (the simulated iron ore body) so that some experience may be gained at tackling a fairly realistic example, to see whether the reader can reproduce the author's result.

Acknowledgements are due to Richard Durham, who provided some of the examples and the simulated iron ore body; Reg Puddy who produced the splendid drawings; Dr C. G. Down who created the situation which forced me to write the book; Malcolm Clark who baby-sat and produced some of the prettier tables; and finally to André Journel and others at Fontainebleau who taught me all I know about the theory of the Theory of Regionalised Variables. Any shortcomings and inaccuracies in the text lie with me.

ISOBEL CLARK

Contents

Symbols used in the Text

h, d, b, l	distances or lengths
$\gamma(h)$	semi-variogram between two points—theoretical model
$\gamma_l(h)$	semi-variogram between two cores along the same borehole—model
$\gamma^*(h)$	experimental point semi-variogram
$\gamma_l^*(h)$	experimental core semi-variogram
$\gamma(l; b)$	semi-variogram between two parallel cores
$\chi(l)$	semi-variogram between a point and a line
$\chi(l; b)$	semi-variogram between a line and a panel
$H(l, b)$	semi-variogram between a point and a panel
$F(l)$	variance of grades within a line
$F(l, b)$	variance of grades within a panel
$F(l, b, d)$	variance of grades within a block
$\bar{\gamma}$	average value of the semi-variogram between two specified 'areas'
a	range of influence of a semi-variogram
C	sill of a semi-variogram
C_0	nugget effect
p	slope of the linear semi-variogram
α	parameter in generalised linear and de Wijsian semi-variograms
g	grade (value)
s	standard deviation of the grades
$\bar{g}$	mean value of the grades
y	logarithm of the grade
$\bar{y}$	mean value of the logarithm of the grade
s_y	standard deviation of the logarithm of the grade

z	Standard Normal deviate
$\phi(z)$	height of the Standard Normal probability density function
$\Phi(z)$	probability that a Standard Normal variate is less than z
c	cutoff value, selection criterion
$\bar{g}_c$	average grade above cutoff
P	proportion of the ore which is above cutoff
A	'area' to be estimated
T	unknown grade of area to be estimated
S	sample set available for estimation
T^*	estimator formed from the sample values
w_i	weight accorded to sample i
σ_ε	standard error of a linear estimator
σ_e	standard error of the arithmetic mean of the samples
σ_k	standard error of kriging estimate
$\delta\theta, \delta h$	a small angle, a small distance

CHAPTER 1

Introduction

Perhaps it would be useful here at the very beginning to clear up any possible ambiguity which arises because of the use of the title Geostatistics. In the early 1960s, after much empirical work by authors in South Africa, Georges Matheron, now Head of the Centre de Morphologie Mathématique in Fontainebleau, France, published his treatise on the Theory of Regionalised Variables. The application of this theory to problems in geology and mining has led to the more popular name Geostatistics. The contents of this book are confined to the simplest application of the Theory of Regionalised Variables, that of producing the 'best' estimation of the unknown value at some location within an ore deposit. This technique is known as kriging. The purpose of this text is to provide a simple treatment of Geostatistics for the reader unfamiliar with the field. The subject may be discussed at a number of levels of mathematical complexity, and it is the intention here to keep the mathematics to a *necessary* minimum. Some previous knowledge must be assumed on the reader's part of basic concepts of ordinary statistics such as mean, variance and standard deviation, confidence intervals and probability distributions. Readers without this background are referred to any one of a large number of excellent basic texts.

The application of Geostatistics to the estimation of ore reserves in mining is probably its most well known use. However, it has been emphasised time and again that the estimation techniques can be used wherever a continuous measure is made on a sample at a particular location in space (or time), i.e., where a sample value is expected to be affected by its position and its relationships with its neighbours. Since most applications—and most of the author's experience—are confined to the mining field, so will most of the examples in this book. Also, there will be a tendency to talk of 'grades' rather than 'sample values', for brevity if nothing else. If the reader is interested in other fields, it suffices to replace 'grade' by porosity, permeability, thickness, elevation, population density, rainfall, temperature, fracture length, abundance or whatever.

The application of statistical methods to ore reserve problems was first attempted some 30 years ago in South Africa. The problem was that of predicting the grades within an area to be mined from a limited number of peripheral samples in development drives in the gold mines. Gold values are notoriously erratic, and when plotted in the form of a histogram show a highly skewed distribution with a very long tail into the rich grades. Normal (Gaussian) statistical theory will not handle such distributions unless a transformation is applied first. H. S. Sichel applied a log-normal distribution to the gold grades and achieved encouraging results. He then published formulae and tables to enable accurate calculation of local averages for log-normal variables, and *also* confidence limits on those local averages. Three major drawbacks exist in the application of Sichel's '*t*' estimator. The 'background' probability distribution must be log-normal. The samples must be independent. There is no consideration taken of the *position* of the samples—all are equally important. However, the technique proved very useful in the gold mines, especially since some measure of the reliability of the estimator was provided. It also laid the base for further statistical work by providing the conceptual framework necessary, i.e., by assuming that the sample values came from some probability distribution. At this stage, it was assumed that all the samples (in a given area) came from the same probability distribution—a log-normal one—and this assumption is known in ordinary statistics as 'stationarity'.

Subsequent to this work attempts were made to incorporate position and spatial relationships into the estimation procedure. Two things seemed sensible: there should be 'rich' areas and 'poor' areas within a deposit; and there should be some sort of relationship between one area and the next. These were tackled in the 1950s and early 1960s by the introduction of Trend Surface Analysis. In South Africa, trends were picked out by forming a 'rolling mean' which produced a smoothed map so that high and low areas could be distinguished. In the United States a 'Polynomial Trend Surface' analysis was propounded which used a statistical technique to fit a mathematical equation to describe the trend. Both methods have one thing in common—the basic assumptions about the statistical characteristics of the deposit. These assumptions have been extended from the 'stationarity' one, by stating that the sample value is *expected* to vary from area to area in the deposit. Some areas are expected to be rich, some to be poor. This expectation can be expressed as a reasonably smooth variation, either by a smoothed map or a relatively simple equation. Round about this trend there is expected to be *random* variation. That is, the value at any point in the deposit is supposed to comprise (i) a 'fixed' component of the trend (which is probably unknown), and (ii) a random variable following one specific distribution. Thus the stationarity has been shifted one step; the expected grade may vary slowly, but the random component is 'stationary'. We have also dropped the log-normality assumption. This approach is quite useful for an overview of the deposit, but, except in heavily sampled areas like the gold mines, is not really useful for local estimation.

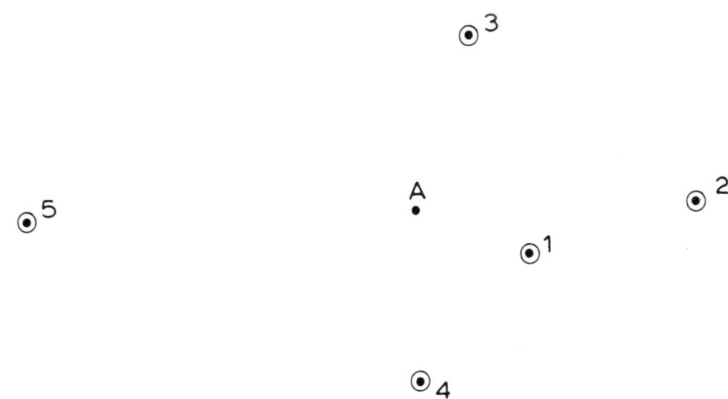

FIG. 1.1. Hypothetical sampling and estimation situation.

Let us consider the problem of local estimation, e.g. of trying to estimate the value at, say, point *A* in Fig. 1.1, given the samples at the various locations shown. It seems reasonable to evolve an estimation procedure which gives more importance to sample 1 than to sample 5. A whole range of methods have been produced to decide on the 'weight' accorded to each sample, mostly based on the distance of the sample from the point being estimated. Sample values may be weighted by inverse-distance, inverse-distance squared, or by some arbitrary constant (e.g. range of influence) minus the distance. All of these involve the same basic assumption—that the relationship between the value at point *A* and any sample value depends on the distance (and possibly direction) between the two positions, and on nothing else. It does not depend on whether one is in a rich or poor zone, or on the actual sample values, but only on the geometric placing of the samples. In fact, it does not even depend on the mineral in the deposit!

There are some problems with this approach. Which weighting factors are the best to choose? How far do you go in including samples—if there is a sample 6 which is twice as far away as sample 5, should it be included? How reliable is the estimate when we get it? Can we seriously expect the same estimation method to be equally valid on all types of deposits? On the other hand, the idea of weighting samples by some measure of their similarity to what is being estimated is intuitively appealing. It also seems to avoid those crippling restrictions on what distribution of values you can handle, which so limit the other methods of estimation. 'Similarity' can be measured statistically by the covariance between samples or by their correlation. However, to calculate either of these we have to go back to 'stationarity' type assumptions. Let us look instead at the *difference* between the samples.

In Fig. 1.1 it seems sound to expect that the value at position 5 will be 'very different' from that at A, whilst sample 1, say, will have a value 'not very different' from that at A. Let us make an assumption that the *difference in value* between two positions in the deposit depends only on the distance between them and their relative orientation. Suppose we took a pair of samples 50 ft apart on a north–south line in one part of the deposit, and measured the difference between the two values. Now, suppose we did the same, say, 200 ft away. And again in yet another position, and so on. The value obtained (difference in grade) would be different for each pair of samples, but under our assumptions all of these values would be from the same probability distribution. Thus, if we could take enough such pairs, we could build up a histogram of the differences and investigate the distribution from which they were drawn. We would expect that distribution to be governed by the distance between the pair and the relative orientation, i.e. 50 ft, north–south. Effectively, we have worked the implicit assumptions of the distance weighting techniques into a statistical form.

However, we will have one histogram for every different distance and direction in the deposit. To build up any useful picture of the deposit we need as many different distances and directions as possible. To investigate a histogram for each would be tedious and would overwhelm us with not terribly useful information. Let us resort to the usual trick of summarising the histogram in a couple of simple parameters. The usual ones are the arithmetic mean (average) and the variance, or equivalently the standard deviation. Suppose, for shorthand, we describe the distance between the samples and the relative orientation as h. We have said that the difference in grade between the two samples depends only on h. In statistical terms, the distribution of the differences depends only on h. If this is true of the whole distribution, it is also true of its mean and variance. That is, we can describe the mean difference in grade as $m(h)$ and the variance of these differences as $2\gamma(h)$. If we had a set of pairs of samples for a specific h (say 50 ft, north–south) then we could calculate an 'experimental' value for $m(h)$:

$$m^*(h) = \frac{1}{n}\sum [g(x) - g(x+h)]$$

where g stands for grade, x denotes the position of one sample in the pair and $x + h$ the position of the other, and n is the number of pairs which we have. You will have noticed the introduction of the '*' to show that this is something we have *calculated* rather than something 'theoretical'. Unfortunately, it can be shown that this is not a very good way of estimating $m(h)$, and that to get a good way involves intense mathematical complications. Let us look closer at $m(h)$ itself. It represents an average difference in grades between two samples—in other words, an 'expected' difference. If $m(h)$ is zero, this implies that we 'expect' no difference between grades a distance h apart. Put another way, we 'expect' the same sort of grades over an area of the deposit which is at least as large as h. In jargon terms, locally (within h) there is no *trend*. It is a convenient assumption to make for our purposes, so we will assume that there is no trend within the scale in which we are interested. We will see later what happens if this is not true.

Having rid ourselves of $m(h)$, let us turn to the variance of the differences. This has been called $2\gamma(h)$ and is usually known as the variogram, since it varies with the distance (and direction) h. In practice, having made our no-trend assumption, we can calculate:

$$2\gamma^*(h) = \frac{1}{n}\sum [g(x) - g(x+h)]^2$$

The '2' in front of the γ is there for mathematical convenience. The term $\gamma(h)$ is called the semi-variogram (although some authors sloppily call it the variogram), and $\gamma^*(h)$ is the experimental semi-variogram; γ^* bears the same relationship to γ that a histogram does to a probability distribution.

Having defined a semi-variogram, what sort of behaviour do we expect it to have. We have a measure of the difference between the grades a distance (and direction) h apart. The measure which we have is in units of grade squared, e.g. (% by weight)^2, (p.p.m.)^2 and so on, and we calculate a value for the experimental semi-variogram for as many different values of h as possible. The easiest way to display these figures is in a graph—hence the name semi-vario*gram*. It is usual to plot the graph as in Fig. 1.2. That is, the distance between the pairs of samples is plotted along the horizontal axis and the value of the semi-variogram along the vertical. By definition h starts

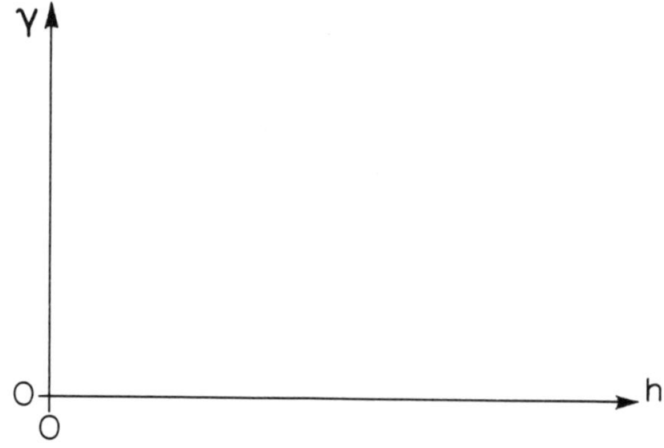

FIG. 1.2. Usual method of plotting a semi-variogram on a graph.

at zero, since it is impossible to take two samples closer than no distance apart. The γ axis also starts at zero, since it is an average of squared values.

Consider the case when h is equal to zero. We take two samples at exactly the same position and measure their values. The difference between the two must be zero, so that γ and γ^* must always pass through the origin of the graph. Now suppose we let the two samples move a little distance apart. We would now expect some difference between the two values, so that the semi-variogram will have some small positive value. As the samples move further apart the differences should rise. In the ideal case when the distance becomes very large the sample values will become independent of one another. The semi-variogram value will then become more or less constant, since it will be calculating the difference between sets of independent samples. This so-called 'ideal shape' for the semi-variogram is shown in Fig. 1.3, and is to Geostatistics as the Normal distribution is to statistics. It is a 'model' semi-variogram and is usually called the spherical or Matheron model. The distance at which samples become independent of one another is denoted by a and is called the range of influence of a sample. The value of γ at which the graph levels off is denoted by C and is called the sill of the semi-variogram. The spherical model is given mathematically as:

$$\gamma(h) = C\left(\frac{3}{2}\frac{h}{a} - \frac{1}{2}\frac{h^3}{a^3}\right) \qquad \text{when } h \le a$$

$$= C \qquad \text{when } h \ge a$$

This model was originally derived on theoretical grounds (as was the Normal distribution) but has been found to be widely applicable in practice.

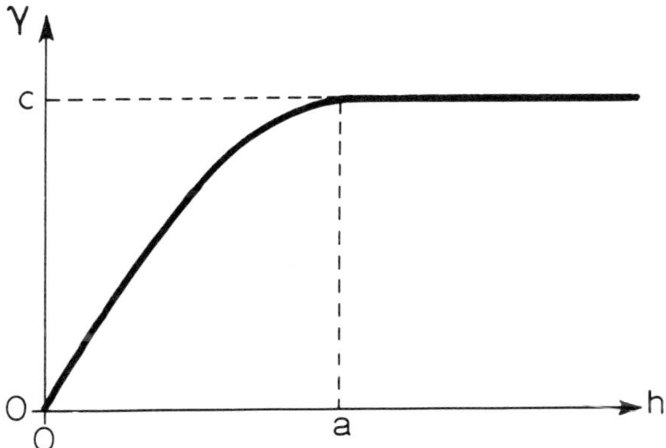

FIG. 1.3. The 'ideal' shape for a semi-variogram—the spherical model.

There are many other possible models of semi-variograms, but only a few are commonly used. One other model with a sill which seems to have found some application is the exponential model. This is described by:

$$\gamma(h) = C[1 - \exp(-h/a)]$$

This model rises more slowly from the origin than the spherical and never quite reaches its sill. Figure 1.4 shows the spherical and exponential with the same range and sill. Figure 1.5 shows the two with the same sill and the same initial slope for comparison. The reason for this will become clear in the next chapter.

One of the interesting properties of models with a sill—both mathematically and for the applications—is that the sill value, C, is equal to the ordinary sample variance of the grades. If you could take a set of random independent observations from the deposit and calculate the sample variance;

$$s^2 = \frac{1}{n-1}\sum(g_i - \bar{g})^2 \qquad \text{where } \bar{g} = \frac{1}{n}\sum g_i$$

then s^2 and C will both be estimates of the same 'true' sample variance. The relationship between s^2 and C will be seen later to have wide-ranging consequences.

There are also models which have no sill. The simplest of these is the linear model:

$$\gamma(h) = ph$$

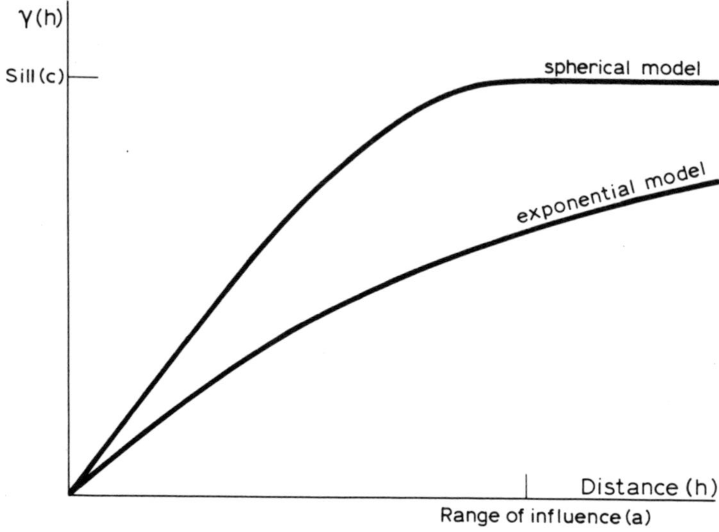

FIG. 1.4. Comparison of the exponential and spherical models with the same range and sill.

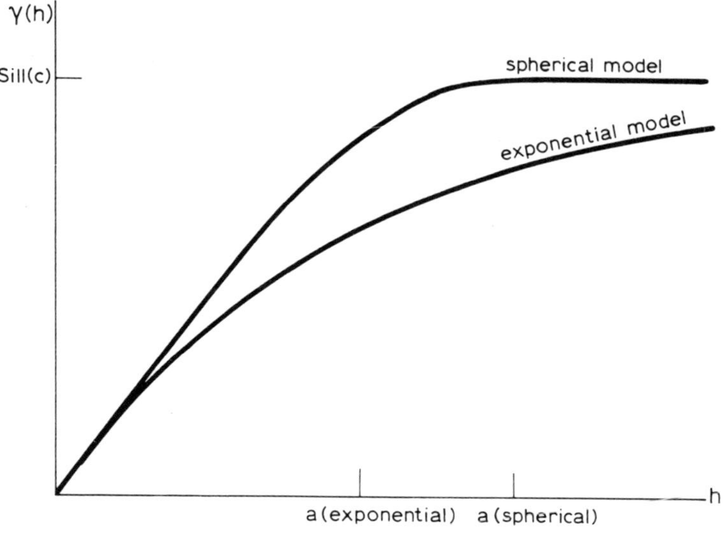

FIG. 1.5. Comparison of the exponential and spherical models with the same initial slope and sill.

where p is the slope of the line. An extension of this model is the 'generalised linear':

$$\gamma(h) = ph^{\alpha}$$

where α lies between 0 and 2 (but must not equal 2). This model is shown in Fig. 1.6 for various values of α. Another model without a sill is the de Wijsian model:

$$\gamma(h) = 3\alpha \log_e(h)$$

in which the semi-variogram is linear if plotted against the logarithm of the distance.

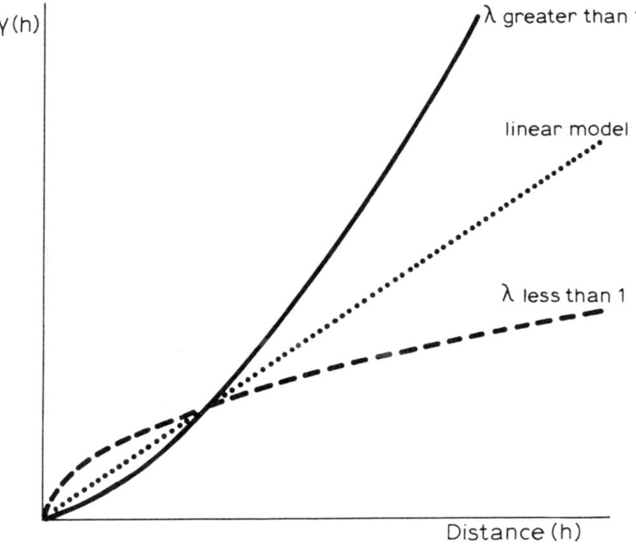

FIG. 1.6. The linear and the generalised linear model $\gamma(h) = ph^{\lambda}$.

One other model exists, to describe the semi-variogram of a purely random phenomenon. Effectively, it is a spherical model with a very small range of influence. The 'nugget effect' as it is called, is given by:

$$\gamma(0) = 0$$
$$\gamma(h) = C_0 \qquad \text{when } h > 0$$

Note that even with completely random phenomena the semi-variogram must be zero at distance zero. Two samples measured at *exactly* the same position must have the same value.

In practice, many semi-variograms comprise a mixture of two or more of these models and we shall see some of these in Chapter 2. To summarise our introduction to Geostatistics, here are the basic assumptions necessary for their application:

(1) Differences between the values of samples are determined only by the relative spatial orientation of those samples.
(2) We are really interested only in the mean and variance of the differences, so our real contention is that these two parameters depend only on the relative orientation. This is known as the 'Intrinsic Hypothesis'.
(3) For convenience we have assumed that there is no trend on the deposit which is likely to affect values within the scale of interest. Thus we are only interested in the variance of the difference in value between the samples.

From these assumptions we have produced the notion of a semi-variogram, and we have discussed the sort of shape which we expect a semi-variogram to take. In the next chapter we will look at the process of actually calculating an experimental semi-variogram and trying to relate it to the models discussed.

CHAPTER 2

The Semi-Variogram

We have seen in Chapter 1 how the definition of a semi-variogram arises out of the notions of 'continuity' and 'relationship due to position within the deposit'. The semi-variogram, γ, is a graph (and/or formula) describing the expected difference in value between pairs of samples with a given relative orientation. We also discussed the ideal forms which semi-variograms might take. We are now going to discuss calculated or 'experimental' semi-variograms.

Consider the data shown in Fig. 2.1. We have here a stratiform iron ore-body, through which a set of drill-holes have been bored, perpendicular to

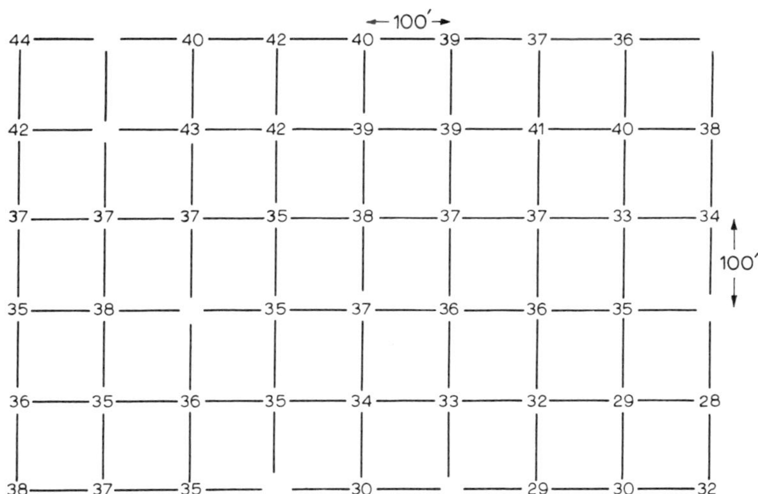

FIG. 2.1. Example of data on a grid for the calculation of an experimental semi-variogram—iron ore.

the dip of the ore. The value given at each location is the average value of Fe (% by weight) over the intersection of the borehole with the ore (see Fig. 2.2). Essentially this is a two-dimensional problem, so that the h in our definition of the semi-variogram depends on the distance between the pair of samples, and their relative orientation in a two-dimensional plane.

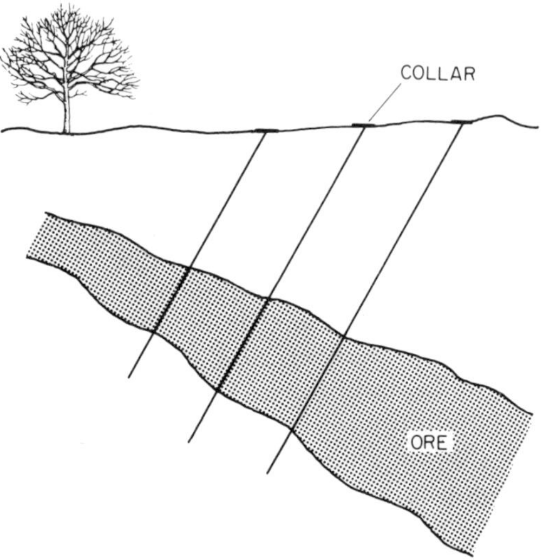

FIG. 2.2. Cross-section through the iron ore deposit.

Let us consider the east–west direction, and try to construct an experimental semi-variogram for this relative orientation. The grid on which the holes have been so conveniently placed is 100 ft by 100 ft, so that we can only calculate values of the experimental semi-variogram, γ^*, for distances which are multiples of this. At zero we know that $\gamma^*(0)$ is equal to zero. At 100 ft we need to find all pairs of samples at a separation of 100 ft in the east–west direction. These are shown in Fig. 2.3. The calculation as defined says: take each pair; measure the difference in value between the two samples; square it; add up all the squares; divide this sum by twice the number of pairs. In our example:

$$\begin{aligned}\gamma^*(100) = [&(40-42)^2 + (42-40)^2 + (40-39)^2 + (39-37)^2 \\ &+ (37-36)^2 + (43-42)^2 + (42-39)^2 + (39-39)^2 \\ &+ (39-41)^2 + (41-40)^2 + (40-38)^2 + (37-37)^2 \\ &+ (37-37)^2 + (37-35)^2 + (35-38)^2 + (38-37)^2\end{aligned}$$

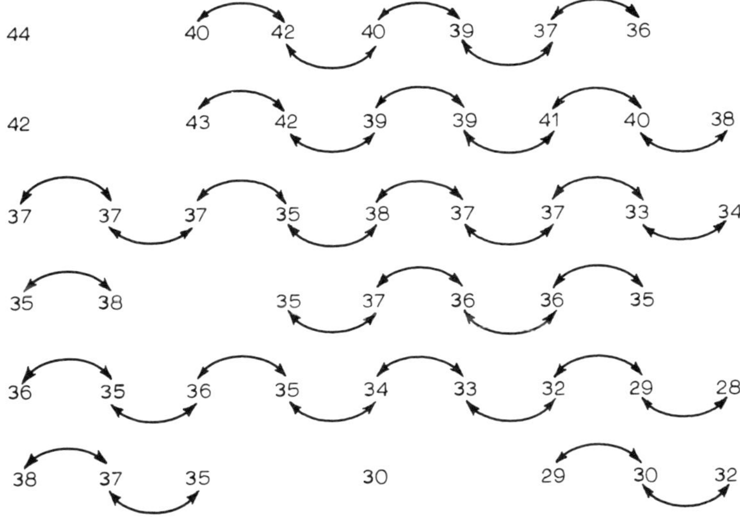

FIG. 2.3. Identifying all the pairs at 100 ft apart in the east–west direction.

$$
\begin{aligned}
&+ (37-37)^2 + (37-33)^2 + (33-34)^2 + (35-38)^2 \\
&+ (35-37)^2 + (37-36)^2 + (36-36)^2 + (36-35)^2 \\
&+ (36-35)^2 + (35-36)^2 + (36-35)^2 + (35-34)^2 \\
&+ (34-33)^2 + (33-32)^2 + (32-29)^2 + (29-28)^2 \\
&+ (38-37)^2 + (37-35)^2 + (29-30)^2 \\
&+ (30-32)^2] \div (2 \times 36)
\end{aligned}
$$

$$\gamma^*(100) = 1{\cdot}46(\%)^2$$

This gives us one point which we can plot on a graph of the experimental semi-variogram (γ^*) versus the distance between the samples (h), that is [100 ft, $1{\cdot}46(\%)^2$]. Now let us consider a distance between samples of 200 ft. Figure 2.4 shows the pairs which lie at this distance in the east–west direction, and the calculation becomes:

$$
\begin{aligned}
\gamma^*(200) = {} & [(44-40)^2 + (40-40)^2 + (42-39)^2 + (40-37)^2 \\
&+ (39-36)^2 + (42-43)^2 + (43-39)^2 + (42-39)^2 \\
&+ (39-41)^2 + (39-40)^2 + (41-38)^2 + (37-37)^2 \\
&+ (37-35)^2 + (37-38)^2 + (35-37)^2 + (38-37)^2 \\
&+ (37-33)^2 + (37-34)^2 + (38-35)^2 + 35-36)^2
\end{aligned}
$$

$$+ (37 - 36)^2 + (36 - 35)^2 + (36 - 36)^2 + (35 - 35)^2$$
$$+ (36 - 34)^2 + (35 - 33)^2 + (34 - 32)^2 + (33 - 29)^2$$
$$+ (32 - 28)^2 + (38 - 35)^2 + (35 - 30)^2 + (30 - 29)^2$$
$$+ (29 - 32)^2] \div (2 \times 33)$$
$$\gamma^*(200) = 3{\cdot}30\,(\%)^2$$

which we can plot on the graph versus 200 ft. The question now arises of where to stop. We could obviously continue up to distances of 800 ft, for

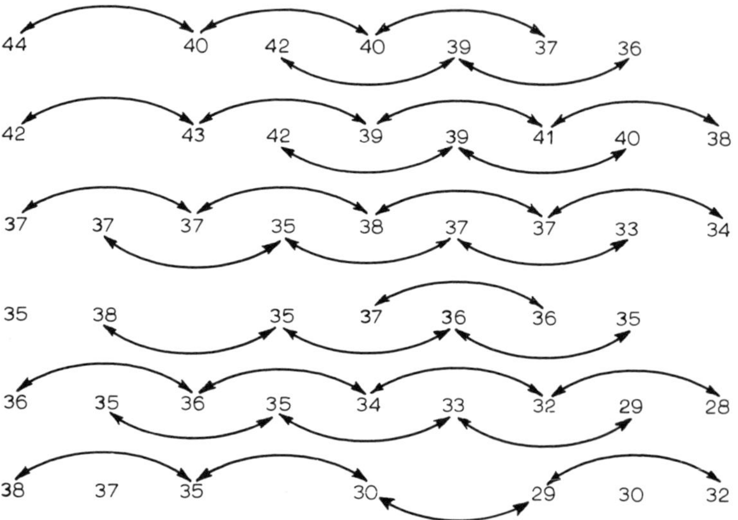

FIG. 2.4. Identifying all the pairs 200 ft apart in the east–west direction.

which we would have 7 pairs. In practice, we rarely go past about half the total sampled extent—in this case, say, 400 ft. Table 2.1 shows the calculated points for the experimental semi-variograms in the east–west and in the north–south direction, and Fig. 2.5 shows a plot of the two γ^*s. There seems to be a distinct difference in the structure in the two directions. The north–south semi-variogram rises much more sharply than the east–west, suggesting a greater continuity in the east–west direction. To verify this, we should then calculate the semi-variogram in at least one 'diagonal' direction, e.g. northwest–southeast. These figures are shown in Table 2.2, and Fig. 2.6 shows the three experimental semi-variograms plotted on the same graph. Of course, the intervals at which the diagonal

TABLE 2.1. CALCULATION OF EXPERIMENTAL SEMI-VARIOGRAM VALUES IN TWO MAJOR DIRECTIONS FOR IRON ORE EXAMPLE ON SQUARE GRID

Direction	*Distance between samples (ft)*	*Experimental semi-variogram*	*Number of pairs*
East–west	100	1·46	36
	200	3·30	33
	300	4·31	27
	400	6·70	23
North–south	100	5·35	36
	200	9·87	27
	300	18·88	21

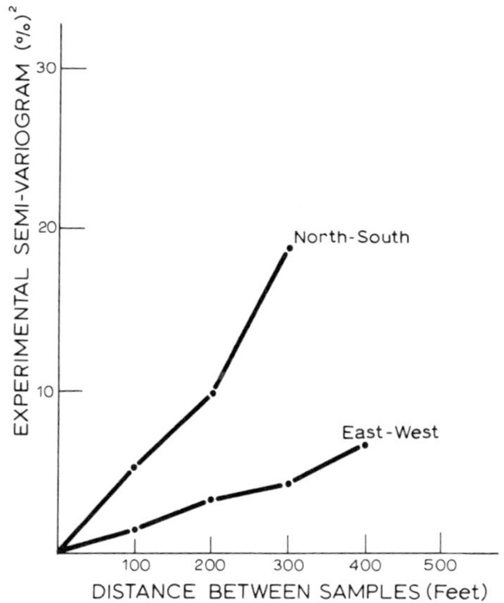

FIG. 2.5. Experimental semi-variograms in the two major directions for the iron ore example.

TABLE 2.2. CALCULATION OF SEMI-VARIOGRAM IN DIAGONAL DIRECTION FOR IRON ORE

Direction	*Distance between samples (ft)*	*Experimental semi-variogram*	*Number of pairs*
North-west	141	7·06	32
South-east	282	12·95	21
diagonal	424	30·85	13

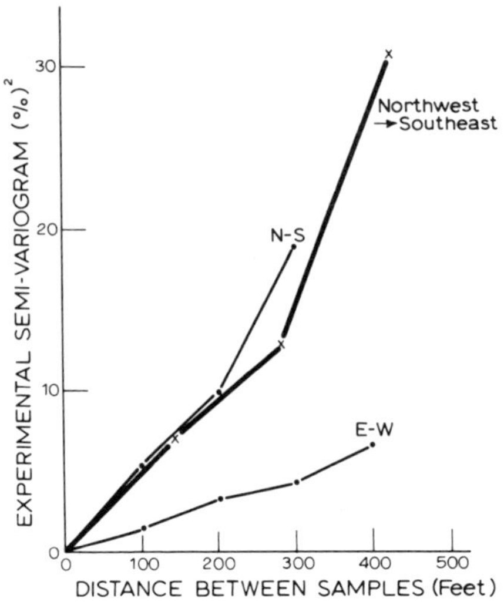

FIG. 2.6. Experimental semi-variograms including a diagonal for the iron ore example.

semi-variogram values are calculated are now multiples of $100\sqrt{2} = 141$ ft. The new γ^* seems to verify the difference between the other two, since it lies between them—although it seems to be closer to the north–south than to the east–west. The conclusion which must be drawn is that more information is needed to determine the 'true' axis of the anisotropy. It would be rather optimistic to suppose that our drill grid was laid down in the exactly correct direction for the different structures. Secondly, we must decide whether, say, the last point on the diagonal semi-variogram is reliable. This was calculated on only 13 pairs, as opposed to the next lowest of 21. Does this mean we should place only two-thirds as much confidence on it? Some theoretical work on simple cases has been done at Fontainebleau, but in practice the only rule is: the fewer pairs, the less reliable. The east–west semi-variogram seems to be reasonably consistent, and suggests a straight line with slope $6{\cdot}5(\%)^2 \div 400\text{ ft} = 0{\cdot}016\,25(\%)^2/\text{ft}$. Thus for the east–west direction:

$$\gamma(h) = 0{\cdot}016\,25h\,(\%)^2$$

For the north–south direction, the following seems reasonable:

$$\gamma(h) = 0{\cdot}05h\,(\%)^2$$

TABLE 2.3. HYPOTHETICAL BOREHOLE LOG FROM LEAD/ZINC DEPOSIT—ZINC VALUES

Depth below collar (m) (top of core)	*Zn* (%)	*Depth below collar (m) (top of core)*	*Zn* (%)
45·40	8·44	98·60	2·56
46·92	6·21	100·12	4·48
48·44	4·01	101·64	8·73
49·96	3·23	103·16	9·64
51·48	2·62	104·68	15·28
53·00	1·20	106·20	core lost
54·52	1·02	107·72	core lost
56·04	0·62	109·24	core lost
57·56	0·20	110·76	core lost
59·08	0·14	112·28	7·56
60·60	0·13	113·80	6·78
62·12	0·24	115·32	7·16
63·64	0·22	116·84	5·51
65·16	0·24	118·36	2·61
66·68	0·22	119·88	3·34
68·20	0·35	121·40	6·80
69·72	0·35	122·92	3·84
71·24	0·34	124·44	3·21
72·76	0·39	125·96	3·90
74·28	0·66	127·48	3·58
75·80	1·40	129·00	4·32
77·32	4·35	130·52	6·00
78·84	7·74	132·04	2·70
80·36	7·06	133·56	3·72
81·88	4·93	135·08	4·80
83.40	3·05	136·60	6·31
84·92	2·42	138·12	7·05
86·44	1·34	139·64	7·24
87·96	0·56	141·16	8·19
89·48	0·53		
91·00	0·70		
92·52	1·01		
94·04	0·95		
95·56	1·20		
97·08	1·87		

That is, in the east–west direction, we 'expect' a squared difference of $0{\cdot}016\,25\,(\%)^2$ for each foot between the samples. Put another way, a difference in grade of $\sqrt{0{\cdot}016\,25} = 0{\cdot}127\,5$ % Fe is expected for two samples 1 ft apart, with a relative orientation of east–west. In the north–south direction the corresponding figure is 0·223 6 % Fe. For samples 100 ft apart, we would expect differences of 1·275 % Fe (east–west) and 2·236 % Fe (north–south) and so on. Thus we have built up a picture of the grade fluctuations within this section of the deposit, and have a fairly simple model to describe the differences in grade.

Now let us turn to another example. Table 2.3 shows a borehole 'log' for one drill hole through a lead/zinc mineralisation which is disseminated in limestone. The first 45·40 m go through barren rock, and the rest of the core has been divided into regular sections 1·52 m (5 ft) long. At one point, the core has been lost—perhaps due to a solution cavity in the limestone. As is the case in most three-dimensional deposits, there is very detailed information 'down' the borehole, but the boreholes are widely scattered over the deposit. The usual practice is to make 'down-the-hole' semi-variograms, and then to look at the horizontal directions as we did in the first example. So, for practice, let us calculate the experimental semi-variogram down this one borehole. Effectively the problem is simpler than the first one since we have one long line of regularly spaced samples with a single gap of 6·08 m. Table 2.4 shows the calculated γ^*, and Fig. 2.7 the plot of this experimental semi-variogram versus the distance between the pairs. In this case the number of pairs of points decreases steadily as the distance increases, from 58 pairs at $h = 1{\cdot}52\,m$ to 28 pairs at $h = 48{\cdot}64$ m. Thus the most 'reliable' points on the graph are those for small distances, and the reliability drops off slowly and regularly. The semi-variogram seems to follow approximately the ideal shape discussed in Chapter 1. It rises from the origin, seems to more or less level off at about 15 m, and continues with some variation around the value, say, of $10{\cdot}5\,(\%)^2$. We could probably fit a spherical model to this semi-variogram without further ado. However, let us look at the supposed variation around the sill. There is a dip in the curve at 25 m, and another at about 35 m. There is *less difference* between samples 25 m apart than there is at 15 m. If we go back to the drill log we find that the grade values seem to rise and fall quite regularly. There is a 'rich' patch centred at about 47 m below collar, another at 81 m and a possible third at 106 m, where the core has been lost. The distances between these rich patches are 34 m and 25 m respectively. Thus the experimental semi-variogram is drawing our attention to the presence of localised rich areas down the borehole. The implications of this would need to be viewed in the

TABLE 2.4. CALCULATED EXPERIMENTAL SEMI-VARIOGRAM FROM LEAD/ZINC DEPOSIT

Distance between samples (m)	*Experimental semi-variogram* $(\%)^2$	*Number of pairs*
1·52	1·33	58
3·04	3·09	56
4·56	5·03	54
6·08	6·70	52
7·60	8·26	51
9·12	9·00	50
10·64	9·67	49
12·16	10·46	48
13·68	11·44	47
15·20	11·87	46
16·72	11·39	45
18·24	11·33	44
19·76	10·93	43
21·28	10·48	42
22·80	9·76	41
24·32	9·21	40
25·84	9·27	39
27·36	11·09	38
28·88	11·70	37
30·40	11·25	36
31·92	9·68	36
33·44	8·60	36
34·96	8·45	36
36·48	9·15	36
38·00	10·15	35
39·52	11·70	34
41·04	13·04	33
42·56	14·03	32
44·08	14·98	31
45·60	15·70	30
47·12	15·94	29
48·64	15·81	28

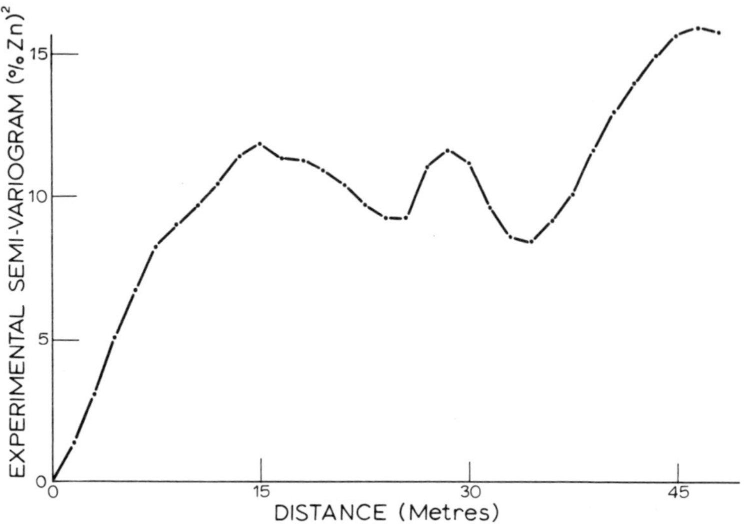

FIG. 2.7. Experimental semi-variogram calculated on one 'borehole' through a hypothetical lead/zinc ore-body.

light of *other* boreholes and/or information about the deposit. If the same sort of pattern occurs on many of the other boreholes then we would suspect some sort of lenticular (or stratified) structure. If the other boreholes do not reflect this regular rise and fall, this is probably just local fluctuation. This particular set of data was taken from a deposit with a marked (geologically) lenticular structure which had already been mapped on-site. This is one manifestation of what happens to the semi-variogram if 'trends'—in this case periodic trends—are present within the deposit and are ignored. On the other hand, for small scale estimation, say up to 20 m in the vertical direction, a spherical model would be quite adequate.

Both of these illustrative examples have been carried out on small sets of data, so that the reader can check his understanding of the calculation by trying to reproduce the answers. The interpretation of an experimental semi-variogram is another matter, and is something that becomes easier with practice. I should like, therefore, to give a few examples of semi-variograms from my own experience.

Table 2.5 shows an experimental semi-variogram which was calculated on silver values from samples taken in a tabular, heavily-disseminated base-metal sulphide deposit. An access adit has been driven into the deposit and a vertical channel sample taken every metre along one wall of the tunnel.

TABLE 2.5. EXPERIMENTAL SEMI-VARIOGRAM FROM 400 m ADIT—SILVER VALUES

Distance between samples (m)	*Experimental semi-variogram*	*Distance between samples (m)*	*Experimental semi-variogram*
1	0·42	51	10·22
2	0·72	52	9·96
3	0·92	53	11·64
4	1·36	54	11·93
5	1·69	55	12·62
6	2·03	56	11·35
7	1·95	57	10·18
8	2·75	58	10·69
9	3·65	59	10·03
10	4·05	60	9·81
11	3·44	61	10·23
12	3·55	62	11·85
13	3·24	63	11·27
14	3·07	64	13·01
15	4·52	65	13·61
16	5·23	66	14·17
17	6·53	67	11·75
18	6·41	68	9·91
19	5·98	69	10·12
20	5·72	70	9·56
21	5·26	71	10·91
22	6·46	72	11·98
23	7·01	73	12·13
24	7·55	74	11·45
25	8·06	75	12·14
26	8·94	76	12·26
27	8·48	77	11·69
28	7·65	78	12·30
29	7·04	79	11·63
30	6·49	80	12·98
31	7·26	81	15·78
32	7·47	82	17·42
33	7·66	83	16·72
34	9·54	84	17·20
35	10·98	85	17·16
36	10·82	86	14·67
37	10·58	87	14·12
38	10·21	88	14·56
39	10·08	89	16·04
40	8·28	90	17·81

TABLE 2.5—*contd.*

Distance between samples (m)	*Experimental semi-variogram*	*Distance between samples (m)*	*Experimental semi-variogram*
41	8·08	91	20·96
42	9·34	92	22·70
43	9·55	93	23·20
44	9·87	94	24·37
45	10·45	95	23·67
46	10·23	96	21·66
47	8·87	97	21·44
48	9·19	98	22·94
49	10·19	99	22·29
50	10·73	100	22·16

Since the width of the ore is variable, the accumulation (grade times width) was calculated for each sample. 400 m of the adit was sampled in this way, giving an unbroken succession of values. The units of accumulation are metres-per-cent (m %), so that the units of the experimental semi-variogram are (m %)^2. Figure 2.8 shows the graph of this semi-variogram versus distance. Near the origin, the points form an almost straight line. This is a

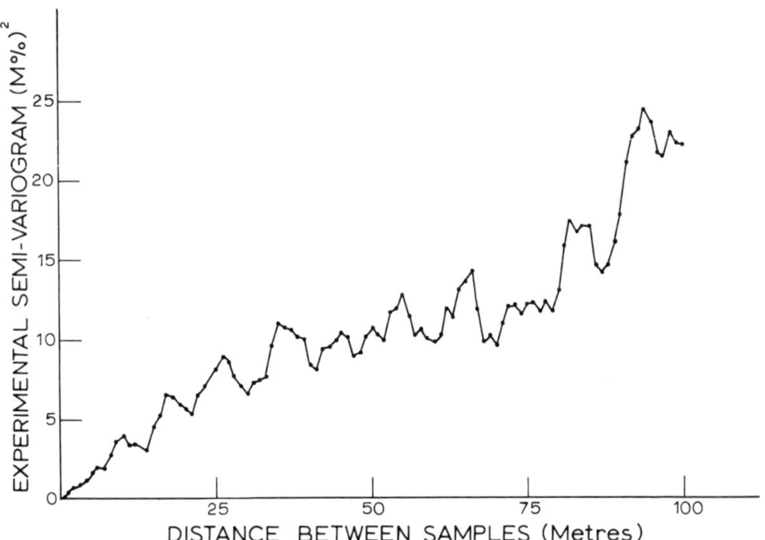

FIG. 2.8. Experimental semi-variogram constructed on the silver values from a complex sulphide deposit.

characteristic of most of the common semi-variogram models. The curve rises, flattens off at about 11 (m %)^2, and then rises again more and more rapidly. In fact, after a distance of about 75 m, the curve is virtually parabolic. *This* is an indication of the presence of a polynomial-type trend within the deposit. There appears to be a smoothly varying large scale trend in operation here. If we wished to consider points more than, say, 75 m apart in any estimation procedure, then we should have to take account of that trend (see Chapter 6). However, if we restrict consideration to areas within the deposit of no more than 75 m in radius, the problem may be safely ignored. Let us, then, look at the semi-variogram only up to distances of 75 m (see Fig. 2.9). A 'sill' appears to exist at $C = 11$(m %)^2. A horizontal

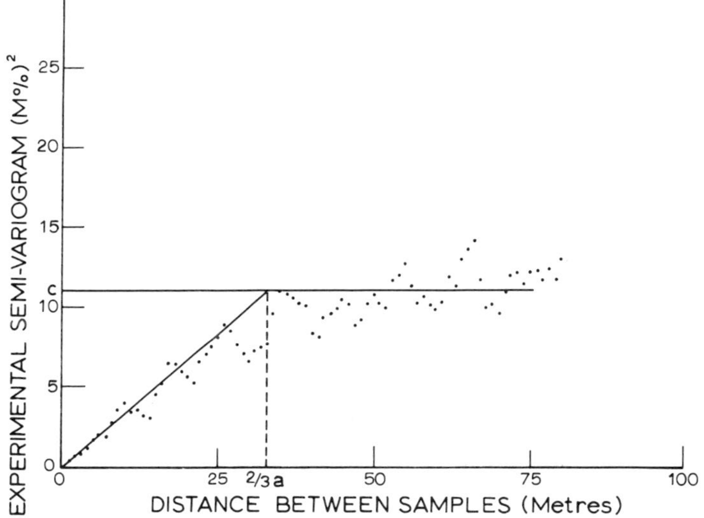

FIG. 2.9. First estimation of model and parameters for the silver semi-variogram.

line has been drawn onto the graph at this level. A more difficult parameter to 'eyeball' is the range of influence a. It can be shown that if a spherical model is to be used—as seems to be indicated by the *flat* nature of the sill—then a line drawn through the first few points of the experimental semi-variogram will intersect the sill at a distance equal to two-thirds of a. Doing this on Fig. 2.9 produces a value of 33 m for the intersection, giving a range of influence of approximately 50 m. Indications are that we need a spherical model with a range of influence of 50 m and a sill of 11 (m %)^2. Since there is no objective (statistical) way of deciding whether a model fits an experimental semi-variogram, the only simple method is to draw the model curve onto the same graph as the experimental one. The equation for this model is:

$$\gamma(h) = 11\left(\frac{3h}{100} - \frac{h^3}{2 \times 50^3}\right) \qquad \text{when } h \leq 50\,\text{m}$$

$$= 11 \qquad \text{when } h \geq 50\,m$$

This curve has been drawn onto the same graph as the experimental points, and the result is shown in Fig. 2.10. The numerical values for various points

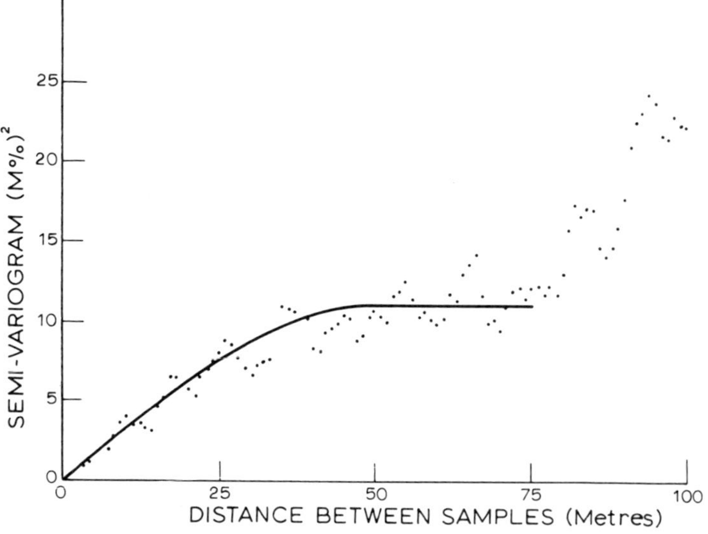

FIG. 2.10. Fitted spherical model to silver semi-variogram.

on the model curve are given in Table 2.6. This seems to give a fairly good fit. It is difficult to see how it might be improved. Sometimes the two parameters require adjustment before an adequate fit is found. Note that the model has only been fitted for distances up to 75 m. Beyond this the trend must be taken into account. In this case we were very lucky, in that the trend does not 'interfere' until after the range of influence is passed. This is not always so, and the closer the parabolic behaviour is to the origin the more heed must be paid to the trend.

It might be argued that a more suitable model for this semi-variogram

TABLE 2.6. SPHERICAL SEMI-VARIOGRAM MODEL FOR SILVER VALUES UP TO $h = 75\,m$

Distance between samples (*m*)	*Theoretical semi-variogram*
0	0·00
5	1·64
10	3·26
15	4·80
20	6·25
25	7·56
30	8·71
35	9·66
40	10·38
45	10·84
50	11·00
>50	11·00

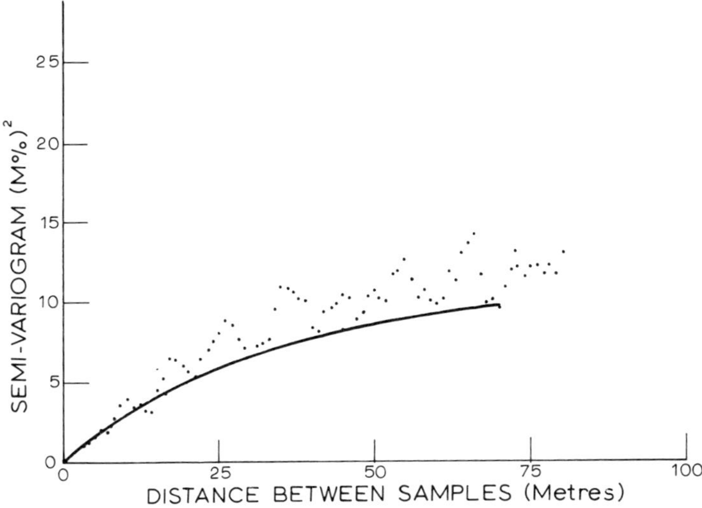

FIG. 2.11. Exponential model with same parameters as fitted spherical (for silver semi-variogram).

would be the exponential model. For interest, let us take the sill again at $11\,(\mathrm{m}\,\%)^2$. For an exponential model the straight line through the origin intersects the sill at a distance *equal* to the range of influence. That is, if we try an exponential model the range will be 33 m. Figure 2.11 shows the model,

$$\gamma(h) = 11[1 - \exp(-h/33)]$$

alongside the data points. The slope at the origin is correct but the rest of the curve is far too low. We can increase the sill to bring up the values, but we also need to increase the value of the range of influence, so that the behaviour near the origin is still correct. Table 2.7 shows the 'model' values given by various sets of parameters—sill and range of influence. Round

TABLE 2.7. ATTEMPTS TO FIT EXPONENTIAL MODELS TO SILVER SEMI-VARIOGRAM

Distance between samples (m)	*Theoretical semi-variograms*			
	$a=33, C=14$	$a=50, C=14$	$a=50, C=15$	$a=50, C=16$
5	1·97	1·33	1·43	1·52
10	3·66	2·54	2·72	2·90
15	5·11	3·63	3·89	4·15
20	6·36	4·62	4·95	5·27
25	7·44	5·51	5·90	6·30
30	8·36	6·32	6·77	7·22
35	9·15	7·05	7·55	8·05
40	9·83	7·71	8·26	8·81
45	10·42	8·31	8·90	9·49
50	10·92	8·85	9·48	10·11
55	11·36	9·34	10·01	10·67
60	11·73	9·78	10·48	11·18
65	12·05	10·18	10·91	11·64
70	12·32	10·55	11·30	12·05

figures have been used for simplicity, but the 'best' exponential fit seems to be the last one, with $a = 50$ m and $C = 16\,(\mathrm{m}\,\%)^2$. Figure 2.12 compares the fit of this curve with the previous spherical model, and with the experimental semi-variogram. I prefer the spherical model because it seems to fit the data between 15 and 40 m better than the exponential. Only a minority of the observed points fall below the exponential curve. A shortening of the range of influence to compensate for this results in a marked change in slope at the beginning of the curve.

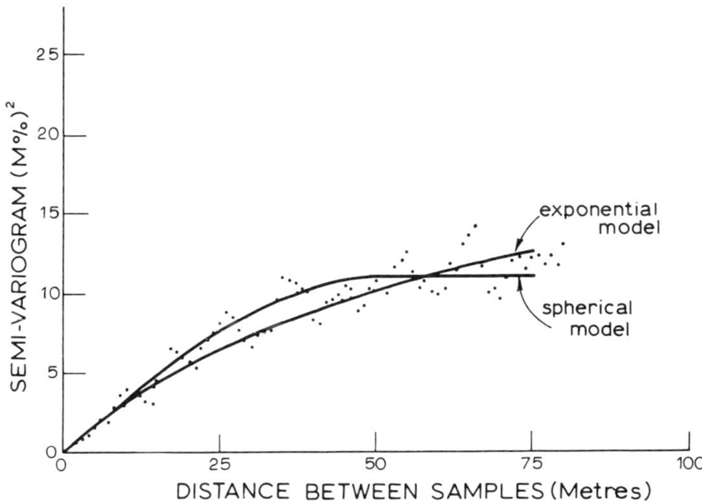

FIG. 2.12. Comparison of final models—exponential and spherical—for silver semi-variogram.

COMPLEX MODELS

Now let us try some real semi-variograms, rather than these hand-picked simple ones. Figure 2.13 shows the experimental semi-variograms for three metals in another complex base-metal sulphide. The metal of economic importance is the copper, but the other two metals are of sufficient value to warrant investigation. The semi-variograms are 'down-the-hole' in direction, and contain information from about 50 boreholes perpendicular to the plane of the ore-body. *My* interpretation of the lead and zinc semi-variograms is pure nugget effect. That is, the 'model' is a horizontal line at a value equal to the sample variance. There appears to be very little relationship even between neighbouring cores! On the other hand, the copper semi-variogram appears to be a combination of a nugget effect (constant) and a parabola. As in the previous example, the parabola implies a polynomial trend, in this case acting on pairs of samples even at 1 m spacing. The nugget effect implies completely random behaviour. So we have a trend with random variation; an ideal case for Trend Surface Analysis.

The next example concerns a nickel ore-body disseminated in peridotite, which has been 'proved' by means of about 45 vertical boreholes. The average spacing between the boreholes was about 60 m and they were not

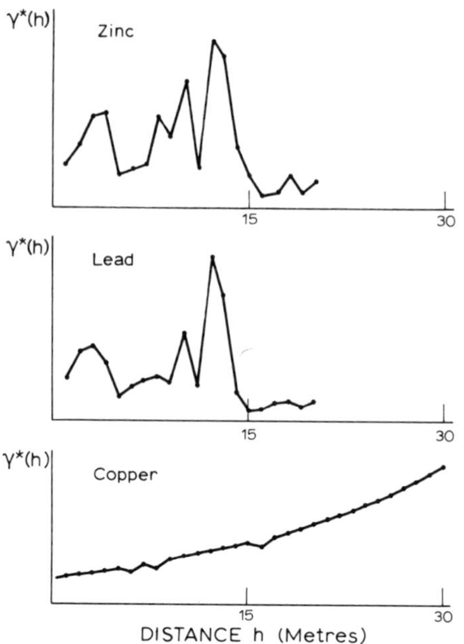

FIG. 2.13. Experimental semi-variograms for a complex base-metal sulphide deposit.

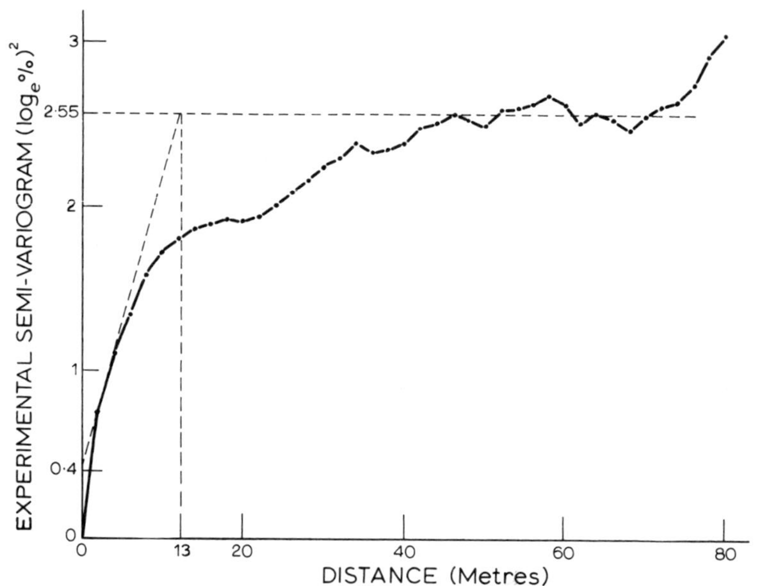

FIG. 2.14. Experimental semi-variogram for a nickel deposit—logarithms of grade values.

regularly spaced, so that only the 'down-the-hole' experimental semi-variograms were calculated. Altogether approximately 4 000 m of core was recovered and assayed in 2 m core sections. In this case the logarithm of the grades was used, rather than the grades; the reason for this has no relevance here. The experimental semi-variogram is shown in Fig. 2.14, and the numerical values are given in Table 2.8. There appears to be a definite flat sill at about 2·55 (log %)^2. However, drawing a straight line through the first two points, as we did in the silver semi-variogram produces two odd results. First, the line intersects the semi-variogram axis at 0·40 (log %)^2 *not* at zero. This suggests that there is a component of each value which is 'random' or unpredictable. Samples very close together still have a reasonably large difference in value. Remembering that the sill (if it exists) is equal to the sample variance, we can see that 0·40 ÷ 2·55 = 0·156 suggests that about 16 % of the variation in the sample values is random and unpredictable. Thus, no matter how closely we sample, this unpredictability will still exist. The semi-variogram model will need to be of the form:

$$\gamma(0) = 0$$
$$\gamma(h) = C_0 + \gamma'(h) \qquad \text{when } h > 0$$

where $\gamma'(h)$ is the usual sort of model (e.g. linear). In effect, the nugget effect is a simple constant raising the whole theoretical semi-variogram 0·4 units. Thus we now seek a model with a sill of 2·15 (log %)^2. Now, we saw in the silver example that extending the initial straight line slope up to the sill gave a value of two-thirds of the range of influence, when using a spherical model. In this case the intersection produces a value of 13 m implying a range of influence of about 20 m. On the other hand, the curve does not even approach the sill until some distance past 45 m. Clearly neither of our ideal models will cope with this sort of situation. Let us look again at the experimental curve. There seems to be an 'intermediate' sill, reached at about 14 m and a value on the γ-axis of 1·95 – 0·40 = 1·55 (to allow for nugget effect). We seem to have a mixture of two spherical type models, one with a shortish range and one with a range of about 50 m. Let us try out this tentative model and see how it fits the experimental semi-variogram. We have a fairly complex model:

$$C_0 = 0{\cdot}40 \text{ (log \%)}^2$$

$$a_1 = 14\text{ m} \qquad C_1 = 1{\cdot}55 \text{ (log \%)}^2$$

$$a_2 = 50\text{ m} \qquad C_2 = 0{\cdot}60 \text{ (log \%)}^2$$

TABLE 2.8. EXPERIMENTAL SEMI-VARIOGRAM FROM A DISSEMINATED NICKEL DEPOSIT (LOGARITHM OF GRADE VALUE)

Distance between samples (m)	*Experimental semi-variogram*	*Number of pairs*
2	0·74	1 222
4	1·10	1 194
6	1·34	1 186
8	1·58	1 152
10	1·72	1 137
12	1·81	1 120
14	1·87	1 095
16	1·90	1 077
18	1·93	1 055
20	1·92	1026
22	1·95	1 011
24	2·01	990
26	2·09	969
28	2·16	950
30	2·25	919
32	2·29	899
34	2·38	886
36	2·35	860
38	2·36	848
40	2·39	825
42	2·48	814
44	2·52	787
46	2·56	779
48	2·55	767
50	2·49	750
52	2·59	736
54	2·61	722
56	2·64	705
58	2·68	689
60	2·62	675
62	2·52	657
64	2·59	639
66	2·53	628
68	2·47	612
70	2·56	597
72	2·62	582
74	2·64	563
76	2·75	552
78	2·93	539
80	3·06	514

Putting these values into the proposed model produces the following:

$$\gamma(h) = 0{\cdot}40 + 1{\cdot}55\left[\frac{3}{2}\frac{h}{14} - \frac{1}{2}\left(\frac{h}{14}\right)^3\right] + 0{\cdot}60\left[\frac{3}{2}\frac{h}{50} - \frac{1}{2}\left(\frac{h}{50}\right)^3\right]$$

for $h \leq 14$ m.

For distances (h) between 14 and 50 m, the model is given by:

$$\gamma(h) = 0{\cdot}40 + 1{\cdot}55 + 0{\cdot}60\left[\frac{3}{2}\frac{h}{50} - \frac{1}{2}\left(\frac{h}{50}\right)^3\right]$$

and when the distance between the two samples is greater than 50 m, the model semi-variogram takes the form:

$$\gamma(h) = 0{\cdot}40 + 1{\cdot}55 + 0{\cdot}60 = 2{\cdot}55$$

In order to compare the theoretical model with the experimental points we

TABLE 2.9. FIRST ATTEMPT TO FIT A MIXTURE OF SPHERICAL MODELS TO THE EXPERIMENTAL NICKEL SEMI-VARIOGRAM (PARAMETERS IN TEXT)

Distance between samples (m)	*Theoretical semi-variogram*
0	0·00
2	0·77
4	1·12
6	1·44
8	1·73
10	1·96
12	2·12
14	2·20
16	2·23
18	2·26
20	2·29
25	2·36
30	2·42
35	2·48
40	2·52
45	2·54
50	2·55
55	2·55
60	2·55

must evaluate the model at various distances, and draw the resulting curve onto the same graph. For example, for distance h equal to 2 m:

$$\gamma(h) = 0{\cdot}40 + 1{\cdot}55\left[\frac{3}{2}\frac{2}{14} - \frac{1}{2}\left(\frac{2}{14}\right)^3\right] + 0{\cdot}60\left[\frac{3}{2}\frac{2}{50} - \frac{1}{2}\left(\frac{2}{50}\right)^3\right]$$

$$= 0{\cdot}766$$

and for a distance h equal to 40 m:

$$\gamma(h) = 0{\cdot}40 + 1{\cdot}55 + 0{\cdot}60\left[\frac{3}{2}\frac{40}{50} - \frac{1}{2}\left(\frac{40}{50}\right)^3\right]$$

$$= 2{\cdot}516$$

A set of values was selected for h and the theoretical curve constructed. The values are shown in Table 2.9, and the resulting model has been plotted in Fig. 2.15. The experimental points are also shown for comparison.

The 'model' curve fits fairly well to the beginning and end of the experimental semi-variogram, but does not seem too good in the middle. The kink in the curve is at far too high a level—it needs to occur at $\gamma = 1{\cdot}95$.

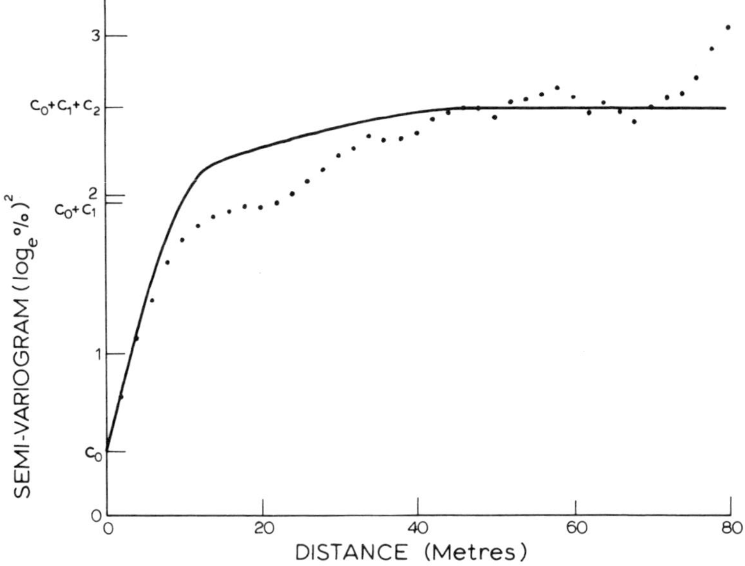

FIG. 2.15. First attempt to fit a mixture of spherical models to the nickel semi-variogram.

We assumed that this level was equal to $C_0 + C_1$. What was forgotten is that, even at short distances, the second spherical component still contributes *some* value to the model, so that the value 1·95 should actually be equal to:

$$C_0 + C_1 + C_2\left[\frac{3}{2}\frac{14}{50} - \frac{1}{2}\left(\frac{14}{50}\right)^3\right]$$

In other words, we need to lower the value of C_1 and raise the value of C_2, and then try the fit again. After a few tries, I got the following model:

$$C_0 = 0{\cdot}40\ (\log\ \%)^2$$

$$a_1 = 12\,\text{m} \qquad C_1 = 1{\cdot}15\ (\log\ \%)^2$$

$$a_2 = 60\,\text{m} \qquad C_2 = 1{\cdot}00\ (\log\ \%)^2$$

This model is shown in Fig. 2.16 alongside the experimental semi-variogram, and seems to be a relatively good fit. Perhaps the reader would like to try to improve upon it? Table 2.10 gives the corresponding numerical values for the model curve. Examples of semi-variogram models which are

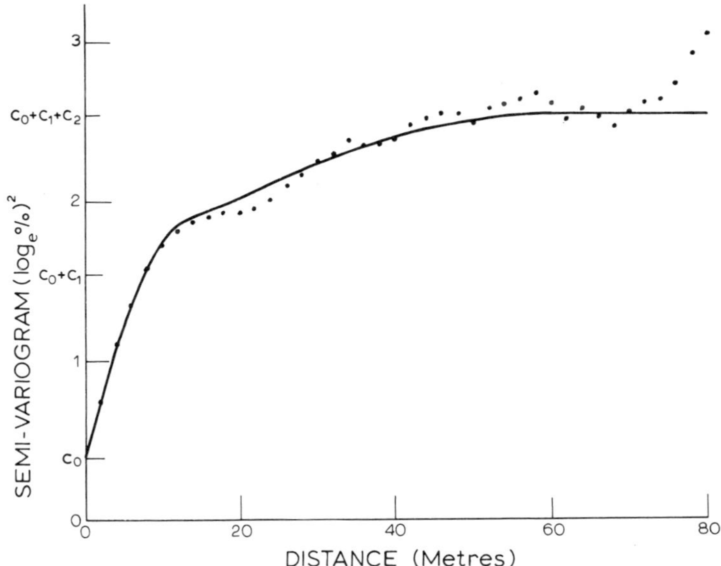

FIG. 2.16. Final attempt to fit a mixture of spherical models to the nickel semi-variogram.

TABLE 2.10. FINAL ATTEMPT TO FIT A MIXTURE OF SPHERICAL MODELS TO THE EXPERIMENTAL NICKEL SEMI-VARIOGRAM (PARAMETERS IN TEXT)

Distance between samples (m)	*Theoretical semi-variogram*
0	0·00
2	0·74
4	1·05
6	1·34
8	1·58
10	1·75
12	1·85
14	1·89
16	1·94
18	1·99
20	2·03
25	2·14
30	2·24
35	2·33
40	2·40
45	2·46
50	2·51
55	2·54
60	2·55

mixtures of spherical components abound in the geostatistical literature, and seem to be about the most common type encountered, especially in low concentration minerals such as cassiterite, copper veins, uranium and so on.

LOG-NORMALITY

I should like, now, to turn to another problem which is often discussed in the literature when samples are expected to follow a log-normal distribution. Whilst the construction of the experimental semi-variogram and the estimation procedures produced by geostatistics do not depend on what distribution the samples follow, there are one or two 'side-effects' which become apparent when dealing with log-normal samples. As every schoolboy knows, the standard deviation of a log-normal distribution is directly proportional to its mean. Consequently the sample variance—and hence the sill of the semi-variogram—is proportional to the square of the mean of the samples. If experimental semi-variograms are constructed on different sets of samples within a deposit, this 'proportional effect' can have a radical effect on the individual experimental semi-variograms. Examples in the literature usually concern cases where, in order to construct experimental semi-variograms in different directions, it has been necessary to use different sets of, e.g. borehole data in each. As an example of 'proportional effect' consider the following situation. In Cornish tin lodes the assay values are usually assumed to follow a log-normal distribution. Such veins are developed by means of horizontal drives approximately 100 ft apart. These drives are sampled every 10 ft by taking chip samples from the roof. In the example under consideration, nine levels have been developed, from 600 ft below surface to 1400 ft (6–14 levels). Semi-variograms were calculated for each level separately. For simplicity, Fig. 2.17 shows only three of these experimental semi-variograms, for levels 6, 10 and 12. The other six lie scattered between levels 6 and 12. Figure 2.18 shows a graph of the average assay value along each drive versus the standard

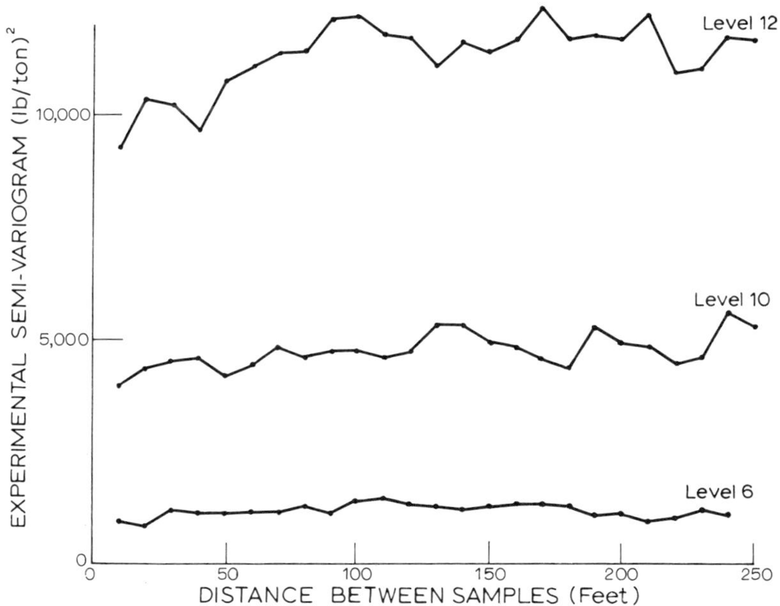

FIG. 2.17. Example of supposed zonal anisotropy—cassiterite vein.

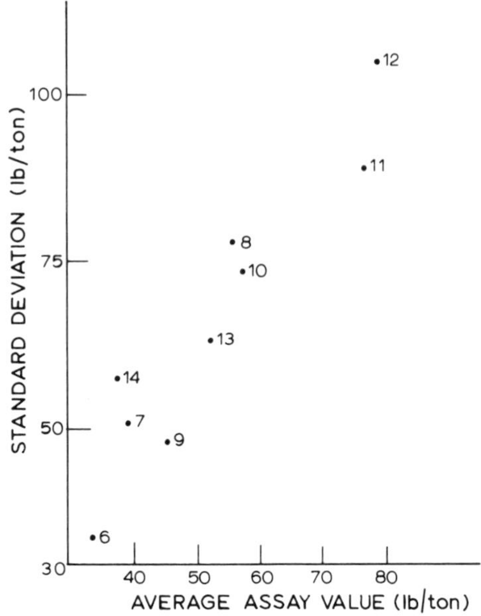

FIG. 2.18. Illustration of the proportional effect—cassiterite.

deviation of the samples along that drive. The averages vary between 35 lb/ton (pounds of SnO_2 per ton of ore) and 80 lb/ton, and the standard deviations vary between 35 and 110 lb/ton. The relationship is virtually perfect between the two, with a calculated correlation coefficient of over 0·85. Since the sill of the semi-variogram is roughly equal to the calculated sample variance, it is easy to see that the experimental semi-variogram for level 6 will be (and is) the lowest, with a sill of about 1200 $(lb/ton)^2$; level 10 will be in the middle with a sill of about 5000; and 12 will be the highest with a sill of 12 000 $(lb/ton)^2$. The question is, can we make an overall semi-variogram for this deposit when the individual experimental semi-variograms vary by an order of magnitude from area to area.

The published authorities state that the valid way to combine these semi-variograms is to 'correct' each one for the proportional effect. That is, to divide the individual experimental semi-variograms by the square of the average of the samples which went into its calculation. This produces a 'relative semi-variogram'—implying that all values given by the semi-variogram are now 'relative to the local mean'. Applying this method to the above example results in nine experimental relative semi-variograms which vary in sill between about 1·00 and 1·80. Notice that these values now have no units. To be converted into meaningful figures they must be multiplied by the square of the local mean. We can now (supposedly) combine these semi-variograms into one for the deposit as a whole, and fit a model to it. If we do so we must remember that in all our estimation procedures etc. we have to 'uncorrect' the values calculated from the semi-variogram—estimation variances, standard errors and so on.

This process of correcting experimental semi-variograms for the proportional effect is widely advocated as the 'right thing to do'. No one seems to have bothered to test whether it actually *works* in practice. In the one case, that described above, where I have been able to investigate in depth and compare what happens if you use the 'relative' semi-variogram, I found that correction by the local mean gave completely erroneous results. Therefore, I would *not* recommend this procedure, but rather that you should combine the original experimental semi-variograms and try to fit a model to the 'uncorrected' data. In the study mentioned above this was found to give the correct values at all times.

OTHER VARIABLES

It has been said time and again that geostatistics—Kriging and so on—can be applied equally well to other variables which are spatially or temporally distributed. This book has been more or less devoted to mining applications, because this is still the major field. However, many other variables can be handled, and I should like to give one or two examples here. Even in mining applications, grade or economic value of the mineral is not always the sole variable of interest. In many deposits the 'thickness' of the deposit is as important as the grade, and in many sedimentary deposits this factor is far more important. In the Cornish tin example described above, the width of the vein is fully as important a variable as the cassiterite content. Both variables are required to assess the economic viability of the lode or portions of it. Figure 2.19 shows the overall experimental semi-variogram calculated for the nine levels, 6–14. To this semi-variogram I fitted a model which consisted of: a small nugget effect, which was slightly surprising; one spherical component with a range of influence of about 30 ft and another with a range of 150 ft.

As an example of other types of spatially distributed data which might be considered, Fig. 20 shows an experimental semi-variogram which was produced during a study of the rainfall characteristics and runoff in a

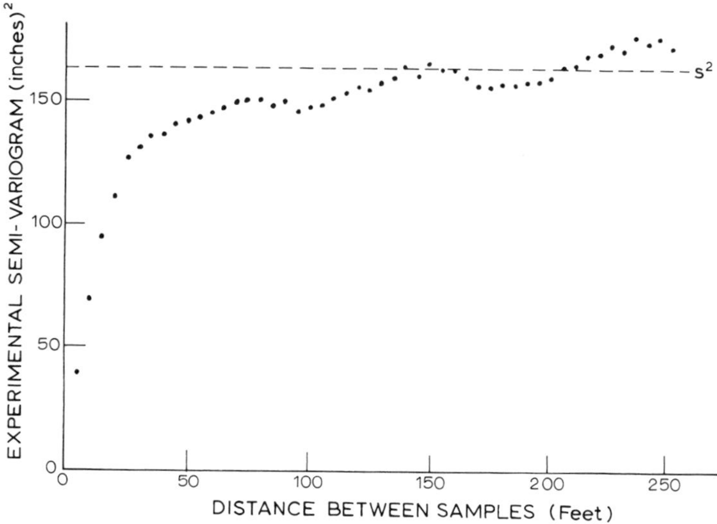

FIG. 2.19. Experimental semi-variogram constructed on the lode widths in the cassiterite vein.

catchment area in the Pennines in England. The data observations are the recorded monthly rainfall figures at rain gauges scattered over the catchment area. The semi-variogram has been constructed without regard to the *direction* of the pair of samples. That is, the author has assumed that his variable shows the same continuity down the long axis (direction of flow of the major river) as it does across the valley. The erroneous nature of this assumption is immediately apparent when given the information that the catchment area measures about 30 km across the valley (short axis). The marked discontinuity in the experimental curve suggests that there is a definite difference between the two major directions. Semi-variograms ought to be constructed for at least two different directions to check for this. The second conclusion which can be drawn from this experimental semi-variogram is that if the same shape is shown by the new individual semi-variograms, then a strong trend is in evidence which must be taken into consideration. When considering the nature of rainfall it does seem sensible to expect different amounts of rain to fall on the tops of mountains than lower down in the valleys. This is a good example of when the 'trend' cannot be ignored in the geostatistical estimation procedure.

And now for a completely different type of application we can take a time series example, rather than one which is spatially distributed. A series of

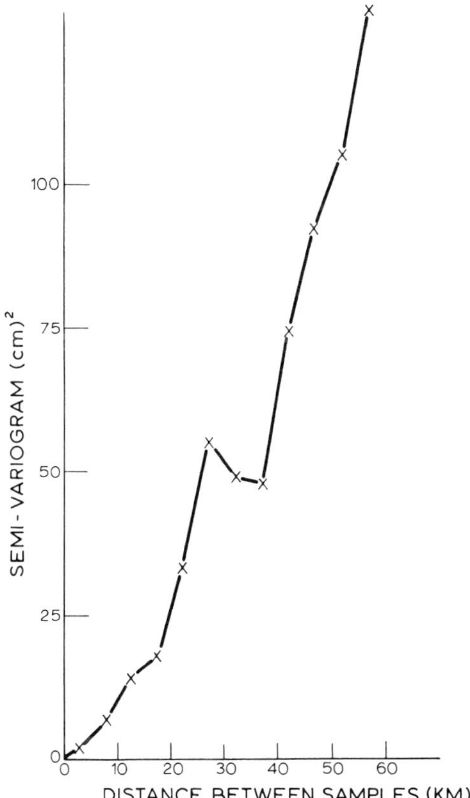

FIG. 2.20. Experimental semi-variogram constructed on the measured rainfall at rain gauge sites.

readings have been taken at the same site in a large river, of various different variables of interest. This is a 'one-dimensional' situation, in which the dimension is time rather than space. Instead of distance between samples, we now have time between samples, so that the horizontal axis of the experimental semi-variogram will now read 'time between observations'. The estimation procedure will be used to predict values of these variables forward into the future, or to fill in gaps in the records caused by machine failure. Figure 2.21 shows two experimental semi-variograms calculated in one case for the temperature of the water, and in the other for the amount of suspended solids contained in the water. The latter looks like an ideal case for a spherical type of model, with a suggestion of a 'trend' at the weekly

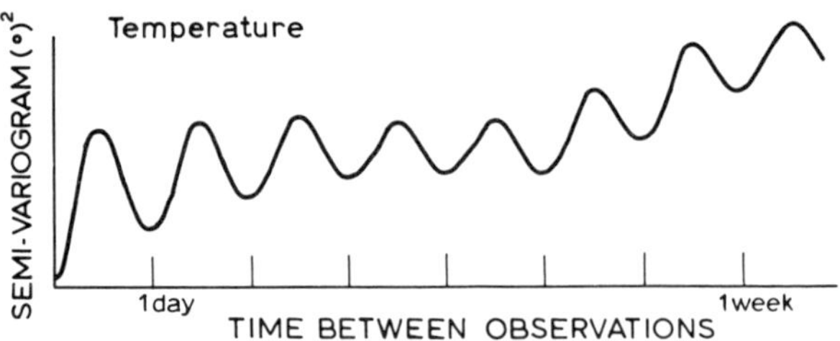

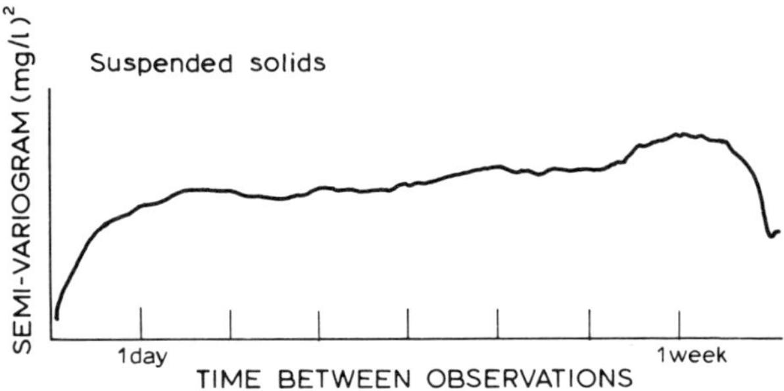

FIG. 2.21. Experimental semi-variograms calculated on water quality variables measured over time.

scale, i.e., fairly homogeneous within any specified week, but varying in level from week to week. The experimental semi-variogram for the temperature shows a perfect daily cycle in temperature, with a little drift coming in after 3 or 4 days.

CONCLUSION

To summarise this chapter, we have seen how to calculate an experimental semi-variogram in one and two dimensions, and how to relate this 'practical' semi-variogram to the 'ideal' models which exist. We have seen that, whilst some deposits may follow fairly simple behaviour, many others require a fairly complex mixture of models to describe the experimental semi-variogram. I have briefly pointed out some problem areas such as strong trends, random phenomena and proportional effect, and tried to indicate how these might be tackled. There are those in authority who say that the fitting of a semi-variogram model is out-moded and unnecessary. To counter this I should like to give an analogy with ordinary statistics. If you take a limited number of samples from an exceedingly large population and construct a histogram, are you prepared to assume that *that* sample histogram describes exactly the behaviour of the whole population? The process of inference—drawing conclusions about the population from a few samples—demands the construction of some sort of model for the behaviour of the whole deposit.

CHAPTER 3

The Volume–Variance Relationship

In the previous chapters we have discussed semi-variograms, calculated experimental semi-variograms and fitted models to these as if the samples had no characteristics other than 'position'. We have ignored the size and shape of the sample, the way in which it may have been taken and/or measured, and so on. We have effectively assumed that the sample values were located at 'points' within the deposit. In this chapter we will see what effect those other characteristics—collectively called 'support'—have on the sample value itself, and hence on the semi-variogram.

Let us consider the lead/zinc example which was discussed in Chapter 2. Although the cores were actually 1·52 m long, we ignored this fact and calculated the experimental semi-variogram as before. Suppose, however, that the cores had been sectioned into 3·04 m lengths instead of 1·52 m—what effect would this have on the sample values and on the semi-variogram? Table 3.1 shows the 'borehole log' for 1·52 and for 3·04 m samples. The experimental semi-variogram was also calculated for the 3·04 m cores, with the results shown in Table 3.2. Figure 3.1 shows the 'new' experimental semi-variogram alongside the 1·52 m one for comparison. For good measure, both tables and the figure also show the resulting values for cores of 4·56 m. It can be seen immediately that the 3·04 m semi-variogram is always lower than the 1·52 m, and the 4·56 m one is considerably lower than both. Let us return to the basic assumptions of Geostatistics and try to explain this behaviour. We must recall two facts from Chapter 1. The first is the basic definition of the semi-variogram: it is the average square of the difference in grade between two samples a given distance apart. If those samples were 'points' then the grade is assumed to be measured 'at a point'; if they are cores then the grade measured is the average grade over the core length. Thus we are not comparing two individual grades, e.g. g_1 and g_2, we are comparing two average grades $\bar{g}_1$

TABLE 3.1. HYPOTHETICAL BOREHOLE LOG FROM LEAD/ZINC DEPOSIT—CORE HAS BEEN SECTIONED IN THREE ALTERNATIVE WAYS, 1·52 m, 3·04 m AND 4·56 m

Depth below collar (m) (top of core)	*1·52 m*	*3·04 m*	*4·56 m*
45·40	8·44	7·32	6·22
46·92	6·21		
48·44	4·01	3·62	
49·96	3·23		2·35
51·48	2·62	1·91	
53·00	1·20		
54·52	1·02	0·82	0·61
56·04	0·62		
57·56	0·20	0·17	
59·08	0·14		0·17
60·60	0·13	0·18	
62·12	0·24		
63·64	0·22	0·23	0·23
65·16	0·24		
66·68	0·22	0·28	
68·20	0·35		0·35
69·72	0·35	0·34	
71·24	0·34		
72·76	0·39	0·52	0·82
74·28	0·66		
75·80	1·40	2·87	
77·32	4·35		6·38
78·84	7·74	7·40	
80·36	7·06		
81·88	4·93	3·99	3·47
83·40	3·05		
84·92	2·42	1·88	
86·44	1·34		0·81
87·96	0·56	0·54	
89·48	0·53		
91·00	0·70	0·85	0·89
92·52	1·01		
94·04	0·95	1·07	
95·56	1·20		1·88
97·08	1·87	2·21	

TABLE 3.1.—*contd.*

Depth below collar (m) (top of core)	1·52 *m*	3·04 *m*	4·56 *m*
98·60	2·56		
100·12	4·48	6·60	7·62
101·64	8·73		
103·16	9·64	12·46	
104·68	15·28		—
106·20	core lost	—	
107·72	core lost		
109·24	core lost	—	—
110·76	core lost		
112·28	7·56	7·17	
113·80	6·78		6·48
115·32	7·16	6·33	
116·84	5·51		
118·36	2·61	2·97	4·25
119·88	3·34		
121·40	6·80	5·32	
122·92	3·84		3·65
124·44	3·21	3·55	
125·96	3·90		
127·48	3·58	3·95	4·63
129·00	4·32		
130·52	6·00	4·35	
132·04	2·70		3·74
133·56	3·72	4·26	
135·08	4·80		
136·60	6·31	6·68	6·87
138·12	7·05		
139·64	7·24	7·71	
141·16	8·19		

and $\bar{g}_2$. We cannot reasonably expect the average grade over 1·52 m of core to have the same behaviour as the grade of a 'teaspoonful' of ore. Similarly, if we take the grade and average it over 3·04 m we would expect different behaviour again. The question is how to characterise this difference in behaviour.

The second fact to recall from Chapter 1 is that the sill of the semi-variogram—if one exists—is equal to the ordinary sample variance. If we

TABLE 3.2. EXPERIMENTAL SEMI-VARIOGRAM VALUES CALCULATED FROM THE THREE DIFFERENT CORE SECTION LENGTHS

Distance between samples (m)	*Experimental semi-variograms* 1·52 *m*	3·04 *m*	4·56 *m*
1·52	1·33		
3·04	3·09	2·67	
4·56	5·03		3·40
6·08	6·70	6·08	
7·60	8·26		
9·12	9·00	8·32	5·91
10·64	9·67		
12·16	10·46	9·50	
13·68	11·44		6·55
15·20	11·87	11·01	
16·72	11·39		
18·24	11·33	10·32	6·68
19·76	10·93		
21·28	10·48	9·18	
22·80	9·76		5·71
24·31	9·21	8·75	
25·84	9·27		
27·36	11·09	10·62	7·21
28·88	11·70		
30·40	11·25	10·10	
31·92	9·68		4·93
33·44	8·60	7·80	
34·96	8·45		
36·48	9·15	8·12	4·28
38·00	10·15		
39·52	11·70	11·55	
41·04	13·04		8·64
42·56	14·03	13·18	
44·08	14·98		
45·60	15·70	14·18	10·01
47·12	15·94		
48·64	15·81	14·20	

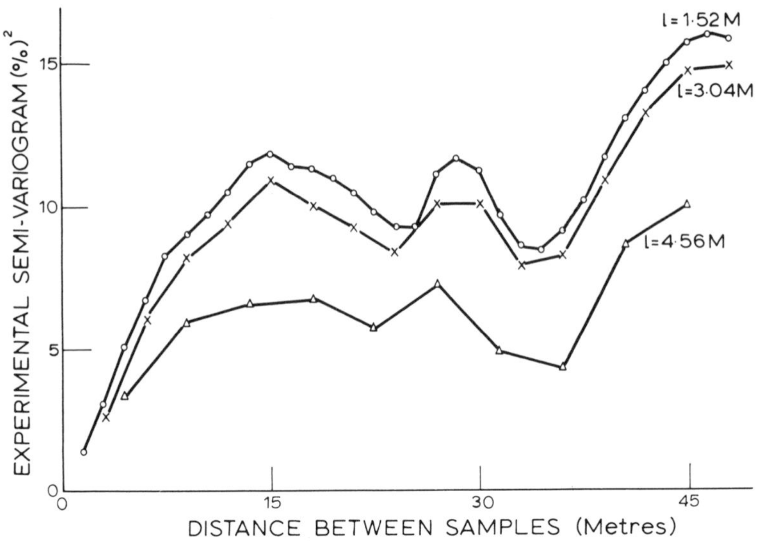

FIG. 3.1. Experimental semi-variograms constructed on various lengths of core—lead/zinc example.

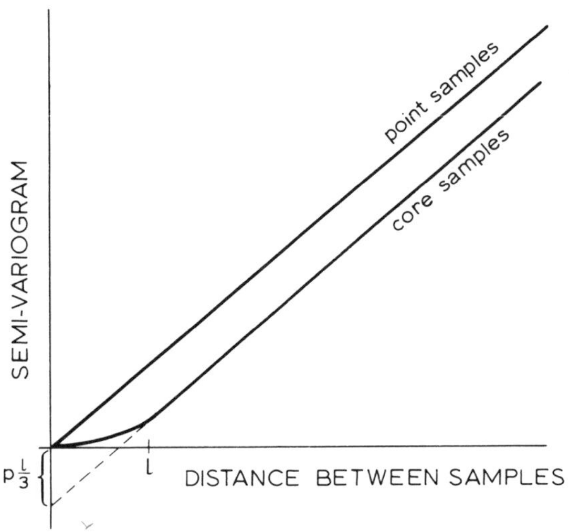

FIG. 3.2. Regularisation of a linear semi-variogram by core lengths.

are dealing with 'point' samples, then we can estimate the sill of the semi-variogram, and compare this value with the sill. That is, $C = s^2$ (ideally). Now, if the samples are cores of a certain length l (e.g. 1·52 m) and we measure the average grade over that length, then we have smoothed out some of the 'point' variation. We have replaced a large number of individual 'points' with one average value. The variance of the averages will therefore be less than the variance of the 'points', so that

$$C_l = s_l^2 < C = s^2$$

In a similar way $C_{3\cdot04}$ will be less than $C_{1\cdot52}$ and so on. If we have a model for the semi-variogram for the point samples we could produce the model for any other size of sample, by employing the mathematical relationship between the point model (γ) and the model for samples of length $l(\gamma_l)$. Since we are only using a limited number of simple models for the point semi-variogram, it is not too difficult to *state* this relationship.

If we have a linear model for the point samples, $\gamma(h) = ph$, where p is the slope of the semi-variogram line, then the semi-variogram for samples of length l is given by:

$$\gamma_l(h) = \frac{ph^2}{3l^2}(3l - h) \qquad \text{when } h \leq l$$

$$= p\left(h - \frac{l}{3}\right) \qquad \text{when } h \geq l$$

This is illustrated in Fig. 3.2, with the point model for comparison. In practice, we generally have an experimental semi-variogram for samples of length l, that is γ_l^*, and we need to find the model for the point samples (γ) for use in the later chapters. Since the slope, p, of the core model is the same as that of the point model, simply measuring the slope of our experimental γ_l^* will give a value for p, and hence the point model, γ. One complication arises if the point model is actually a linear model plus a nugget effect. Taking core samples lowers the line, but a nugget effect will raise it again. From the above formula, if no nugget effect is present, extending the line of the core model *back* until it intersects the semi-variogram axis should produce an intercept of $-pl/3$. Once an estimate of p has been made this can easily be checked, and if necessary a nugget effect C_0 added to the model.

Now suppose our deposit followed an exponential model, with sill C for 'point' samples, i.e.,

$$\gamma(h) = C[1 - \exp(-h/a)] \qquad \text{when } h \geq 0$$

For cores of length l, the theoretical model becomes:

$$\gamma_l(h) = C\left\{2\frac{a}{l} + \frac{a^2}{l^2}[1 - \exp(-l/a)]\{\exp(-h/a)[1 - \exp(l/a) - 2]\}\right\}$$

when $h \geq l$

with a rather more complex form for distances less than the length of the core ($h < l$). Since we are unlikely to have values of an experimental semi-variogram for distances less than the sample length, the form of it seems rather academic. Figure 3.3 shows a point exponential model and the corresponding 'regularised' curve for a sample of length l. It can easily be seen that C_l is lower than C. In fact:

$$C_l = 2C\left\{\frac{a}{l} - \frac{a^2}{l^2}[1 - \exp(-l/a)]\right\}$$

so that a sample which was, say, one-fifth of the range of influence, would produce a sill:

$$C_l = 2C[5 - 25(1 - e^{-0 \cdot 2})] = 0{\cdot}94\,C$$

That is, the new sill will only be 94 % as high as that of the point model. It will also be noticed from Fig. 3.3 that extending the 'linear' part of the core model (close to the origin) until it intersects the sill produces an estimate of the range of influence for the cores a_l, which is *longer* than that of the

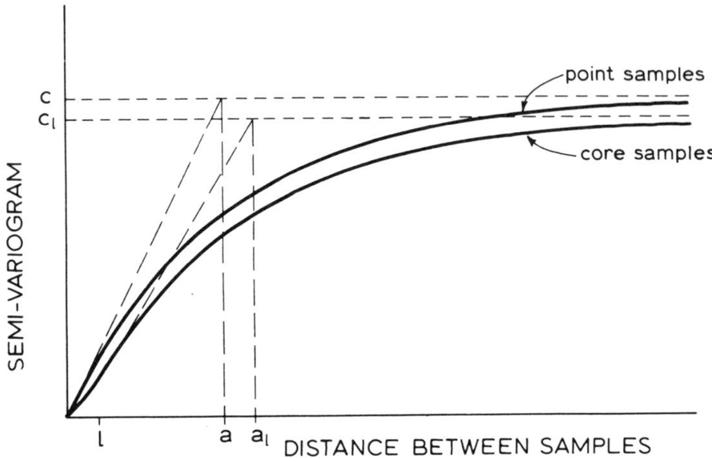

FIG. 3.3. Regularisation of an exponential semi-variogram by core lengths.

points, a. In fact, $a_l = a + l$. This seems quite sensible if you remember that *cores* will have to be just that bit further apart before they become independent.

The above arguments and formulae apply to the situation where you know the 'point' model and you wish to find the 'regularised' model. In practice the situation is generally reversed. We usually have an experimental semi-variogram which has been calculated on cores of a given length, and we need to find the point model for use in the estimation techniques. Suppose, then, that we have a graph of the experimental semi-variogram γ_l^*, and we have decided that our deposit follows an exponential model. The first step is to guess the two parameters C_l and a_l. Since the model is exponential, the sill C_l will be greater than most of the experimental points on the graph. Having guessed C_l, produce a line up through the first two or three points on the graph until it cuts the sill. This will give a first guess at a_l. We know that $a = a_l - l$, so we have a first estimate of a. Using this in the above formula for C_l, we can reverse the equation and produce a value for C, the point sill. We now have guesses at the values of a and C which govern the point model. The next question is whether these are 'good' guesses. We have already stated that if we know the point model, we can produce the corresponding model for cores of any given length, i.e. $\gamma_l(h)$. If our guesses are good ones then this theoretical model for $\gamma_l(h)$ should match the experimental semi-variogram, $\gamma_l^*(h)$. Substituting values for h, l, a and C produces a smooth curve like the lower one in Fig. 3.3, and this can be compared to the data. If necessary, a and C can be altered until the 'model' values become a good fit to the 'data' values. In effect, this is the same procedure as was used in Chapter 2, but with an additional consideration of the sample length.

Let us now turn to the most common model—the spherical model. This will be influenced in the same sort of way as the exponential. The sill for the cores will be lower than that for the 'points', and:

$$C_l = \frac{C}{20}\left(20 - 10\frac{l}{a} + \frac{l^3}{a^3}\right) \qquad \text{for } l \leq a$$

and

$$C_l = \frac{C}{20}\frac{a}{l}\left(15 - 4\frac{a}{l}\right) \qquad \text{for } l \geq a$$

The formula for the semi-variogram of the cores is extremely complex because of the 'discontinuity' in the model but an example is shown in Fig. 3.4. A subroutine to evaluate the formula has been published. If the calculations are to be done by hand (or hand calculator) then it is easier to

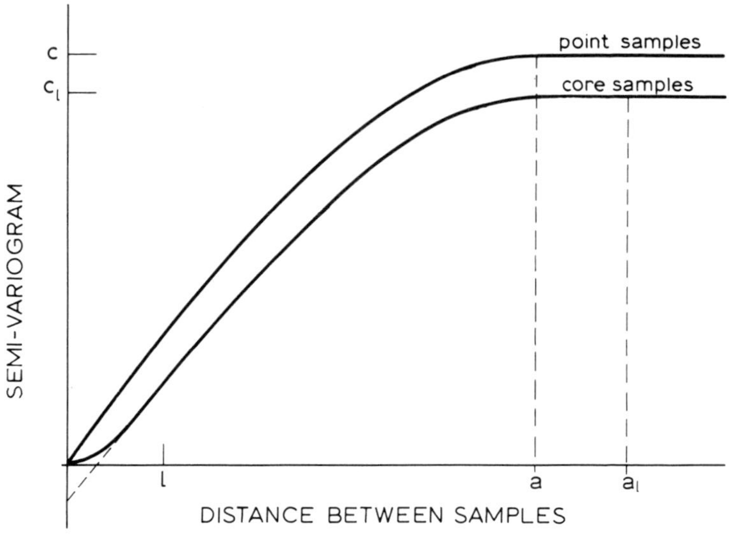

FIG. 3.4. Regularisation of a spherical semi-variogram by core lengths.

use tables such as Table 3.3. This table shows the form of the 'regularised' semi-variogram for a core of length l if the original point semi-variogram had a range of influence a, and a sill of 1. The use of this table is best illustrated by an example. We can now return to the example shown in Fig. 3.1 of the zinc values measured over core lengths of 1·52 m. In Chapter 2 we guessed that the sill lay at about 10·5 (%)^2. This is our first approximation of C_l. Producing the line through the first two points on the experimental semi-variogram gives $2a_l/3 = 9{\cdot}6$ m (approximately). That is, $a_l = 14{\cdot}4$ m, and hence $a = 12{\cdot}9$ m. Using the formula:

$$C_l = \frac{C}{20}\left(20 - 10\frac{l}{a} + \frac{l^3}{a^3}\right)$$

$$10{\cdot}5 = \frac{C}{20}\left(20 - 10\frac{1{\cdot}52}{12{\cdot}9} + \frac{1{\cdot}52^3}{12{\cdot}9^3}\right)$$

$$10{\cdot}5 = 0{\cdot}9412\,C$$

$$C = 11{\cdot}2\,(\%)^2$$

The first estimates, then, for the parameters of the point model are $a = 12{\cdot}9$ *m* and $C = 11{\cdot}2$ (%)^2. We must find the row in the Table 3.3 which corresponds to our value of a/l, i.e. 8·5. The entries along this line

TABLE 3.3. REGULARISED SEMI-VARIOGRAM $\gamma_l(h)$ FOR SPHERICAL MODEL WITH RANGE a AND SILL 1·0 FOR VARIOUS DISTANCES h

a/L	h/L 1.0	2.0	3.0	4.0	5.0	6.0	7.0	8.0	9.0	10.0
.50	.300	.325	.325	.325	.325	.325	.325	.325	.325	.325
1.00	.450	.550	.550	.550	.550	.550	.550	.550	.550	.550
1.50	.463	.678	.681	.681	.681	.681	.681	.681	.681	.681
2.00	.412	.728	.756	.756	.756	.756	.756	.756	.756	.756
2.50	.355	.717	.802	.803	.803	.803	.803	.803	.803	.803
3.00	.307	.669	.822	.835	.835	.835	.835	.835	.835	.835
3.50	.269	.610	.812	.858	.858	.858	.858	.858	.858	.858
4.00	.239	.555	.778	.868	.876	.876	.876	.876	.876	.876
4.50	.215	.507	.733	.861	.889	.889	.889	.889	.889	.889
5.00	.194	.464	.686	.836	.896	.900	.900	.900	.900	.900
5.50	.178	.428	.642	.802	.890	.909	.909	.909	.909	.909
6.00	.163	.396	.601	.764	.872	.914	.917	.917	.917	.917
6.50	.151	.368	.564	.726	.845	.909	.923	.923	.923	.923
7.00	.141	.344	.530	.690	.814	.895	.926	.929	.929	.929
7.50	.132	.323	.500	.655	.782	.874	.923	.933	.933	.933
8.00	.124	.304	.472	.623	.751	.849	.912	.936	.938	.938
8.50	.117	.287	.447	.593	.720	.822	.894	.933	.941	.941
9.00	.110	.272	.425	.566	.690	.794	.874	.924	.943	.945
9.50	.104	.258	.404	.541	.663	.767	.851	.910	.941	.947
10.00	.099	.246	.386	.517	.636	.741	.827	.892	.933	.949

correspond to multiples of the sample length l. That is, $h/l = 1$ means $h = 1{\cdot}52$ m, $h/l = 2$ means $h = 3{\cdot}04$ m and so on. We see that at $h/l = 1$ the table gives a value of 0·116. This would be for a semi-variogram with a sill of 1. Since we have a sill of $11{\cdot}2(\%)^2$, the value we require is $0{\cdot}116 \times 11{\cdot}2 = 1{\cdot}30(\%)^2$. This is now a 'model' value for the semi-variogram of cores of length 1·52 m and can be plotted on the graph next to the 'observed' value of $1{\cdot}33(\%)^2$.

A second point on the model would be at $h = 3{\cdot}04$ m, i.e. $h/l = 2$. The table gives a value of 0·288 for $C = 1$, so that our model value is $0{\cdot}288 \times 11{\cdot}2 = 3{\cdot}23(\%)^2$. This can be compared with the experimental value of $3{\cdot}09(\%)^2$. This process is repeated until we have a model value to compare with each observed value. The resulting model curve has been plotted in Fig. 3.5. This seems to be rather a good fit to the experimental

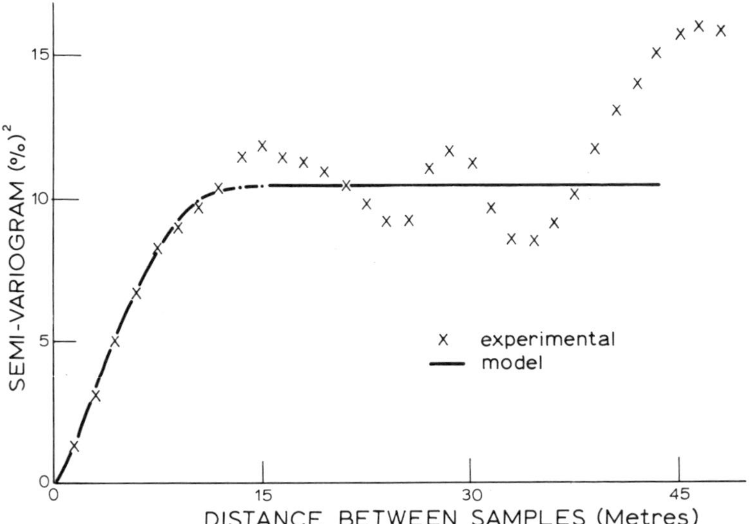

FIG. 3.5. Fitted regularised model to the lead/zinc example—1·52 m cores.

semi-variogram, if we accept the sill at $10{\cdot}5(\%)^2$. Adjustments could be made if the sill was thought to be too low, by raising C and a. Suppose we accept this 'point' model with $a = 12{\cdot}9$ m and $C = 11{\cdot}2(\%)^2$. We can run a secondary check by comparing the models for core lengths 3·04 m and 4·56 m. For the former, $a/l = 4.25$ so that we must interpolate in the table between $a/l = 4{\cdot}00$ and $a/l = 4{\cdot}50$. Linear interpolation is generally sufficient for this sort of exercise. Figure 3.6 shows the experimental and

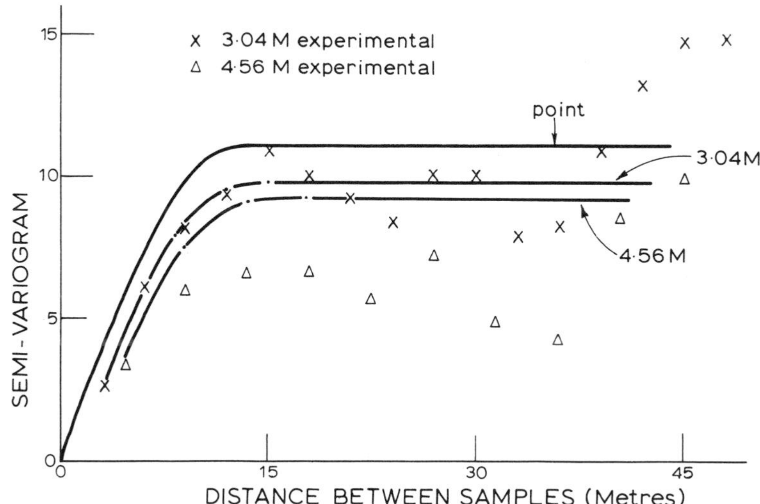

FIG. 3.6. Fitted models to the lead/zinc example—3·04 m and 4·56 m cores and the 'point' model.

model curves for each sample length, and the point model for comparison. The model seems to be a good fit to the 3·04 m semi-variogram, especially to the first four points. However, after the first point on the 4·56 m semi-variogram the model here is consistently considerably higher than the experimental semi-variogram until *h* is about 41 m. This could perhaps be neglected in view of the fact that each of these experimental values is calculated on 15 or fewer pairs. All in all, the spherical model as estimated seems to be a pretty good fit.

VOLUME–VARIANCE CALCULATIONS

This process of the semi-variogram changing with different 'support' is usually known in the literature as 'regularisation'—on the basis that the samples get more regular as the sample size increases. We have seen that we can handle experimental semi-variograms for core samples, and still derive the supposed point model. However, this leads us on to another problem of the 'volume-variance' relationship and the influence of sample size on the sort of distribution encountered. Suppose at the pre-feasibility stage of investigating a deposit the management requests a grade/tonnage

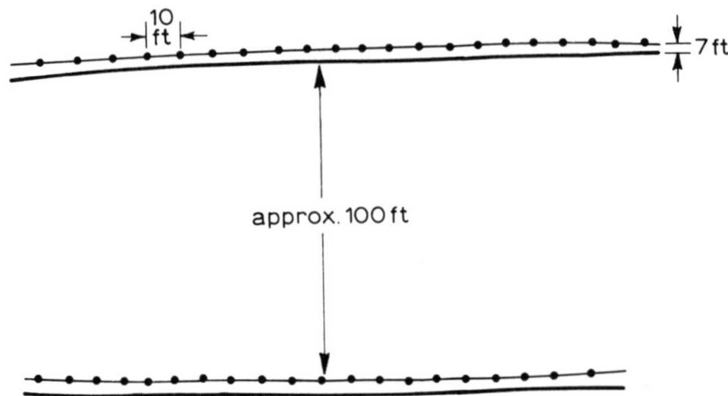

FIG. 3.7. Typical sampling situation in Cornish tin example.

calculation. That is, given an economic cutoff grade (or list thereof) can we evaluate (i) the tonnage of ore in the deposit which is above cutoff and (ii) the average grade of that ore. Suppose we take an example to illustrate the problem which arises. A hydrothermal tin vein has been sampled by means of nine development drives approximately 100 ft apart in the plane of the lode. Chip samples are taken every 10 ft along these drives. The sampling setup is shown in Fig. 3.7. These chip samples may be considered as 'points' since they have a very small volume. Figure 3.8 shows a histogram of the 2730 chip samples taken from the development drives in this lode. Suppose we now specify a 'cutoff grade' of 25 lb/ton for this lode. The histogram shows that about 44 % of the chip samples lie below 25 lb/ton. We could

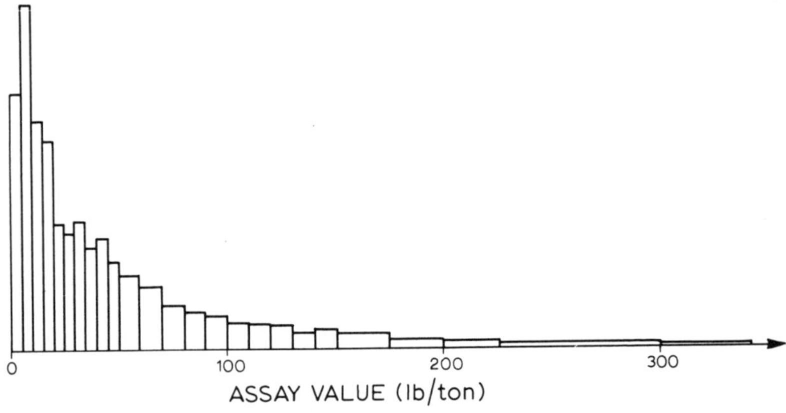

FIG. 3.8. Histogram of chip samples taken from the drives in the cassiterite vein.

(possibly) make the statement that we therefore believe that 44 % of the ore in the lode lies below 25 lb/ton. Now, the usual method of estimating the value in the stopes is to delineate a block (say 125 ft long) between the drives and allocate to that block the average of all the peripheral development samples. It is this estimate which determines whether a stope block enters 'reserves' or not. Figure 3.9 shows the corresponding histogram of the estimates of 125 ft by 100 ft stoping blocks, i.e. the averages of the drive samples over two lengths of 125 ft each. We have seen from the previous

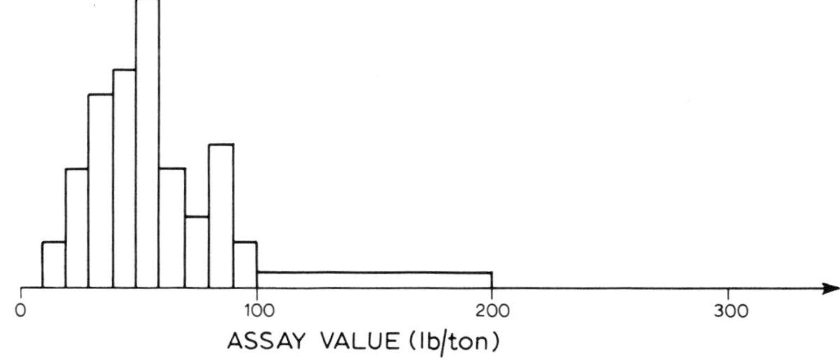

FIG. 3.9. Histogram of estimates of stope values in the cassiterite vein.

exercise that we expect averages over lengths to be somewhat less variable than 'point' samples. This is adequately borne out by the behaviour of these estimates. Whereas the point values range up to 300 lb/ton or more, the drive averages seldom exceed about 150 lb/ton. Whilst 44 % of the point samples lie below 25 lb/ton something less than 8·5 % of the 'block estimates' do so. Should we now say that 8·5 % of the ore lies below 25 lb/ton? What we really need to do is to redefine the phrase 'of the ore'. In the first case what we meant was that if the deposit were divided into chip samples, we could reject 44 % of these as being below cutoff. In the second it was 8·5 % of the drive averages below cutoff. That is, if the deposit were divided into pairs of 125-ft strips 100 ft apart, 8·5 % of these would be below cutoff. Or alternatively, *by my estimate* 8·5 % of the stope blocks would be below cutoff. In other words, we cannot define how much ore we have after selection *unless* we define a unit of selection in terms of size and shape. The real question is 'how many 125 by 100 ft stope panels are below cutoff?' To answer this question we must determine what sort of distribution these panels would follow. The full answer will depend on (i) the distribution of the original samples and (ii) the semi-variogram of the deposit.

Let us make a general statement of the problem and see how it leads to a solution. The original sample data has a 'support' of, say, l; it has a semi-variogram $\gamma_l(h)$ with a sill C_l; it has a distribution of grades which can to some extent be characterised by the histogram and which has a mean $\bar{g}_l$ and variance C_l. The panels or blocks being estimated will have a support of, say, v; a semi-variogram $\gamma_v(h)$ with sill C_v; a distribution with mean $\bar{g}_v$ and variance C_v. The first thing we can say is that $\bar{g}_l$ and $\bar{g}_v$ should be the same, since both describe the average grade of ore over the whole deposit. Thus we can replace them both by $\bar{g}$, the average of 'point' samples. The second thing we can say is that if we have a model for the point semi-variogram we can state the relationship between the point sill C and the 'core' sill C_l, and between C and C_v for any defined volume v. Suppose we take the simple example of a core of length l which can be represented as a straight line

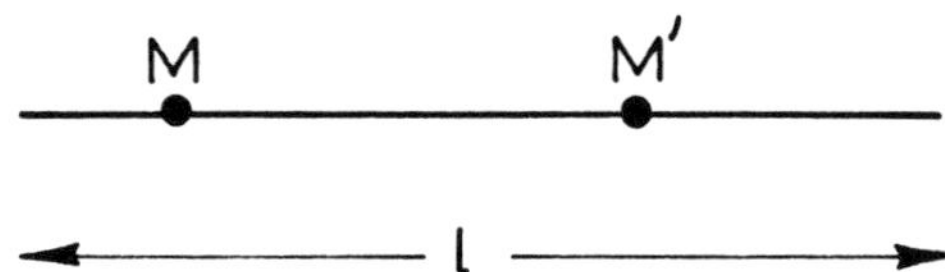

FIG. 3.10. Derivation of the variance of grades within a 'line' segment.

(since the diameter is very much smaller than the length). This is illustrated in Fig. 3.10. Consider two points on this line, M and M′. We could calculate from the model semi-variogram the 'difference' between the grades at these two points. Now suppose we took all possible pairs (M, M′) which exist within the line—including the case when M = M′. In this way we could get a measure of the 'variability' of the grades within the line. If we take the average of the semi-variogram values $\gamma(\mathrm{M} - \mathrm{M}')$ over all possible pairs,

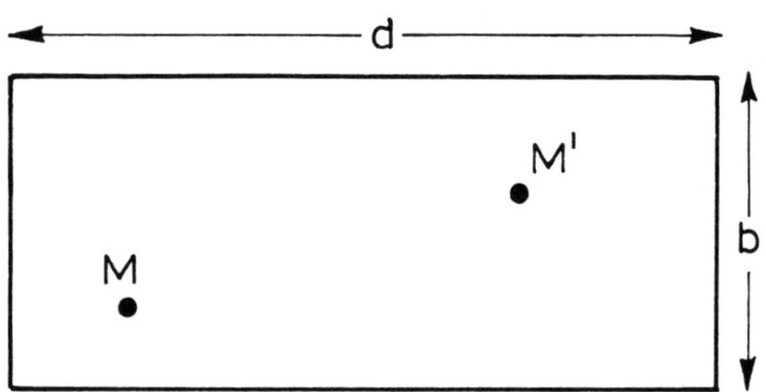

FIG. 3.11. Derivation of the variance of grades within a panel.

then we obtain the variance of the grades *within* the length l. This is the variance which is removed from the system if we only consider the average grade over the length l, i.e. the difference between the point sill and the regularised sill, $C - C_l$. Mathematically:

$$F(l) = \frac{1}{l^2}\int_0^l \int_0^l \gamma(\mathrm{M} - \mathrm{M}')\,\mathrm{d}\mathrm{M}\,\mathrm{d}\mathrm{M}'$$

where $F(l)$ defines the variance of grades within the length l. Although this looks fearsome, it reduces to:

$$F(l) = \frac{pl}{3} \quad \text{for the linear semi-variogram}$$

$$= C\left\{1 - 2\frac{a}{l} + \frac{a^2}{l^2}[1 - \exp(-l/a)]\right\}$$

for the exponential semi-variogram

and for the spherical model:

$$F(l) = \frac{C}{20}\frac{l}{a}\left(10 - \frac{l^2}{a^2}\right) \qquad \text{when } l \leq a$$

$$= \frac{C}{20}\left(20 - 15\frac{a}{l} + 4\frac{a^2}{l^2}\right) \qquad \text{when } l \geq a$$

These, of course, correspond exactly with the difference between the point and regularised semi-variograms. Now suppose we want to consider a two-dimensional panel such as that shown in Fig. 3.11. The F function now becomes $F(d, b)$ to show that it has two dimensions. This would be a quadruple integral, since the points M and M′ can now move throughout the whole panel. The formulae get complicated, but not impossible, and for example of the type of values encountered, Table 3.4 has been produced. This table shows the $F(d, b)$ function for a spherical model with range equal to 1 and sill of 1. This is a 'standardised' spherical model—in the same sense as a 'Standard' Normal distribution. This table can be used to produce the corresponding value of the F function for *any* spherical model, as follows:

(i) divide the lengths of the sides of the panel by the range of influence a;
(ii) read off the corresponding entry in the table;
(iii) multiply this value by C.

TABLE 3.4. AUXILIARY FUNCTION F(L, B) FOR SPHERICAL MODEL WITH RANGE 1·0 AND SILL 1·0.

L \ B	.1	.2	.3	.4	.5	.6	.7	.8	.9	1.0
.10	.078	.120	.165	.211	.256	.300	.342	.383	.422	.457
.20	.120	.155	.196	.237	.280	.321	.362	.401	.438	.473
.30	.165	.196	.231	.270	.309	.349	.387	.424	.460	.493
.40	.211	.237	.270	.305	.342	.379	.415	.451	.484	.516
.50	.256	.280	.309	.342	.376	.411	.445	.479	.511	.541
.60	.300	.321	.349	.379	.411	.443	.476	.507	.538	.566
.70	.342	.362	.387	.415	.445	.476	.506	.536	.565	.591
.80	.383	.401	.424	.451	.479	.507	.536	.564	.591	.616
.90	.422	.438	.460	.484	.511	.538	.565	.591	.616	.640
1.00	.457	.473	.493	.516	.541	.566	.591	.616	.640	.662
1.20	.520	.534	.551	.572	.593	.616	.638	.660	.682	.701
1.40	.572	.584	.600	.618	.637	.657	.677	.697	.716	.733
1.60	.614	.625	.639	.655	.673	.691	.709	.727	.744	.760
1.80	.650	.659	.672	.687	.703	.719	.736	.752	.767	.782
2.00	.679	.688	.700	.713	.728	.743	.758	.773	.787	.800
2.50	.735	.743	.752	.763	.775	.788	.800	.813	.824	.835
3.00	.775	.781	.789	.799	.809	.820	.830	.841	.851	.860
3.50	.804	.810	.817	.825	.834	.843	.852	.861	.870	.878
4.00	.827	.832	.838	.845	.853	.861	.870	.878	.885	.892
5.00	.860	.864	.869	.874	.881	.887	.894	.901	.907	.913

L \ B	1.2	1.4	1.6	1.8	2.0	2.5	3.0	3.5	4.0	5.0
.10	.520	.572	.614	.650	.679	.735	.775	.804	.827	.860
.20	.534	.584	.625	.659	.688	.743	.781	.810	.832	.864
.30	.551	.600	.639	.672	.700	.752	.789	.817	.838	.869
.40	.572	.618	.655	.687	.713	.763	.799	.825	.845	.874
.50	.593	.637	.673	.703	.728	.775	.809	.834	.853	.881
.60	.616	.657	.691	.719	.743	.788	.820	.843	.861	.887
.70	.638	.677	.709	.736	.758	.800	.830	.852	.870	.894
.80	.660	.697	.727	.752	.773	.813	.841	.861	.878	.901
.90	.682	.716	.744	.767	.787	.824	.851	.870	.885	.907
1.00	.701	.733	.760	.782	.800	.835	.860	.878	.892	.913
1.20	.736	.764	.788	.807	.823	.854	.876	.892	.905	.923
1.40	.764	.790	.811	.828	.842	.870	.890	.904	.915	.931
1.60	.788	.811	.829	.845	.858	.883	.901	.914	.924	.938
1.80	.807	.828	.845	.859	.871	.894	.910	.921	.931	.944
2.00	.823	.842	.858	.871	.882	.903	.917	.928	.936	.948
2.50	.854	.870	.883	.894	.903	.920	.932	.941	.948	.957
3.00	.876	.890	.901	.910	.917	.932	.942	.950	.955	.964
3.50	.892	.904	.914	.921	.928	.941	.950	.956	.961	.969
4.00	.905	.915	.924	.931	.936	.948	.955	.961	.966	.972

L \ B	.1	.2	.3	.4	.5	.6	.7	.8	.9	1.0
.10	.099	.136	.178	.222	.266	.309	.350	.390	.428	.464
.20	.168	.196	.231	.269	.308	.347	.385	.423	.458	.491
.30	.239	.262	.291	.324	.358	.394	.429	.463	.496	.527
.40	.311	.329	.353	.382	.413	.445	.476	.508	.538	.566
.50	.380	.395	.416	.441	.468	.497	.526	.554	.581	.607
.60	.445	.459	.477	.499	.523	.549	.574	.600	.624	.648
.70	.507	.519	.535	.554	.576	.598	.622	.644	.666	.687
.80	.565	.574	.588	.606	.625	.645	.666	.686	.705	.724
.90	.616	.625	.637	.652	.669	.687	.706	.724	.741	.757
1.00	.662	.669	.680	.694	.709	.725	.741	.757	.772	.786
1.20	.735	.741	.750	.760	.772	.785	.797	.810	.822	.833
1.40	.789	.794	.800	.809	.818	.828	.839	.849	.858	.867
1.60	.828	.832	.838	.845	.852	.861	.869	.877	.885	.892
1.80	.858	.861	.866	.872	.878	.885	.892	.899	.905	.911
2.00	.880	.883	.887	.892	.897	.903	.909	.915	.920	.925
2.50	.918	.920	.923	.926	.930	.934	.938	.942	.946	.949
3.00	.940	.941	.944	.946	.949	.952	.955	.958	.960	.963
3.50	.954	.955	.957	.959	.961	.963	.966	.968	.970	.972
4.00	.963	.964	.965	.967	.969	.970	.972	.974	.976	.977
5.00	.974	.975	.976	.978	.979	.980	.981	.983	.984	.985

L \ B	1.2	1.4	1.6	1.8	2.0	2.5	3.0	3.5	4.0	5.0
.10	.526	.577	.619	.653	.683	.738	.777	.807	.829	.861
.20	.550	.598	.638	.671	.699	.751	.789	.816	.837	.868
.30	.581	.626	.663	.693	.719	.768	.803	.829	.849	.877
.40	.616	.657	.691	.719	.742	.787	.819	.843	.861	.887
.50	.652	.689	.720	.745	.767	.808	.836	.858	.874	.898
.60	.688	.722	.749	.772	.791	.828	.854	.873	.887	.909
.70	.723	.753	.777	.798	.815	.847	.870	.887	.900	.919
.80	.756	.782	.804	.822	.837	.865	.886	.901	.912	.929
.90	.785	.809	.828	.843	.857	.882	.900	.913	.923	.937
1.00	.811	.832	.849	.862	.874	.896	.912	.923	.932	.945
1.20	.853	.869	.882	.893	.902	.919	.931	.940	.947	.957
1.40	.883	.896	.906	.915	.922	.936	.945	.952	.958	.966
1.60	.905	.915	.924	.931	.937	.948	.956	.961	.966	.972
1.80	.922	.930	.937	.943	.948	.957	.963	.968	.972	.977
2.00	.934	.941	.947	.952	.956	.964	.969	.973	.976	.981
2.50	.955	.960	.964	.967	.970	.975	.979	.982	.984	.987
3.00	.967	.971	.974	.976	.978	.982	.985	.987	.988	.991
3.50	.975	.978	.980	.982	.983	.986	.988	.990	.991	.993
4.00	.980	.982	.984	.986	.987	.989	.991	.992	.993	.994
5.00	.987	.988	.989	.990	.991	.993	.994	.995	.995	.996

Examples of such calculations are given later in this chapter. Similar tables may be produced for the linear and exponential models.

In three dimensions the problem of calculating the $F(l, b, d)$ function analytically appears to be insurmountable. It is necessary to resort to a numerical approximation using a computer. The easiest way to do this is to go back to the definition of the F function: we take pairs of points (M, M′) within the block; consider *all* such pairs; calculate the semi-variogram value between M and M′; sum all these values and average them—this gives the F value. Now, suppose we do not take *all* of the pairs but only a few 'representative' ones. That is, instead of considering the block as an infinite number of points we consider it to be a 'grid' containing a finite number of points, say on a 5 by 5 by 5 grid. Some authors suggest taking 'randomly' distributed points, but there seems little sense in that. Using such a method, Table 3.5 was produced for the 'standardised' spherical model. In order to produce only one table, it has been necessary to insist that two sides of the block have the same length. This table is used in the same way as the two-dimensional one.

GRADE/TONNAGE CURVES

So, we now know how to calculate the function F for one, two and three dimensions, and hence can state the difference between the 'point' variance and the 'regularised' variance of regular shaped areas and volumes. This will give us a numerical quantity for the reduction in the variance, but unless we make some assumptions about the distribution of the samples, we cannot actually quantify the change in the 'tonnage above cutoff' and so on. There are two ways to approach the problem:

(i) Assume that the histogram of the samples represents the whole deposit accurately.

(ii) Assume that the histogram represents a set of samples from the whole deposit, and as such contains some random variation from the 'population' distribution.

The first approach declares that the samples are 'typical' of the whole deposit, and leads to graphical anamorphisms and transfer functions. The second approach declares the belief that if we could measure the grade at every point in the deposit we would end up with a smooth curve of a fairly simple form. This is a much simpler approach, and generally seems to be sufficient for most deposits.

To start with a simple example, let us consider an iron ore deposit which is known to follow a Normal distribution with a mean of 48 % Fe and a standard deviation of 5 % Fe. This distribution has been established on samples small enough to be called 'points'. We also know that the deposit follows a point semi-variogram model which is spherical with a range of influence of 400 ft. Now, suppose that the mine plan is to be constructed on blocks which are 100 ft by 100 ft by 50 ft. What will the distribution of these blocks look like. The first thing we can say is that it will probably be Normally distributed. It will certainly have the same mean (48 % Fe) as the 'points'. The only change will be in the standard deviation of the distribution. We need to evaluate the function $F(100, 100, 50)$ for a spherical model with $a = 400$ and $C = 25$. To use Table 3.5 we must 'standardise' the situation so that the range of influence becomes 1. That is, $F(100, 100, 50)$ for $a = 400$ is the same as $F(0{\cdot}25, 0{\cdot}25, 0{\cdot}125)$ for $a = 1$. Table 3.5 gives a value of 0·209 for these arguments, but this is for a model whose sill is 1. For our model the required value is $0{\cdot}209 \times 25 = 5{\cdot}225\,(\%\,\text{Fe})^2$. This is the difference between the point variance and the block variance. Therefore the variance of the block values will be $25 - 5{\cdot}225 = 19{\cdot}775\,(\%)^2$ leading to a block standard deviation, s_v, of 4·45 % Fe. This is slightly over 10 % less than the point standard deviation, as would be expected with such a 'small' block. Thus we have two distributions to be considered, both Normal, as follows:

(i) point distribution: $\bar{g} = 48\,\%$ Fe; $s = 5\,\%$ Fe
(ii) block distribution: $\bar{g} = 48\,\%$ Fe; $s_v = 4{\cdot}45\,\%$ Fe

These two distributions are shown in Fig. 3.12, and the reduction in spread

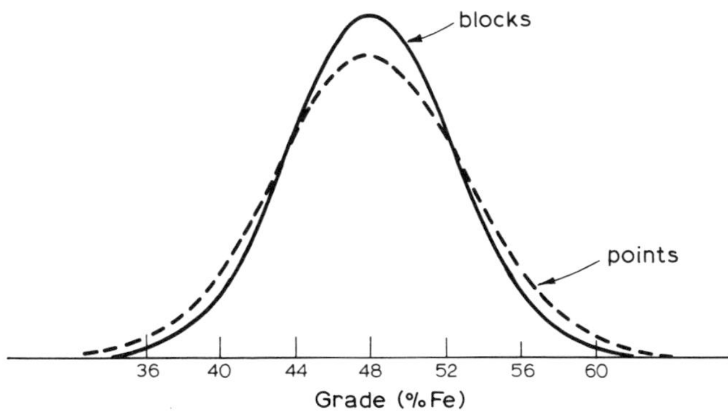

FIG. 3.12. Comparison of the distributions of points and blocks within a hypothetical iron ore deposit.

for the block distribution can be clearly seen. To see how the difference in the distributions will affect the grade–tonnage calculations, let us take as an example a cutoff grade of 44 % Fe. The proportion (P) of the distribution which is above cutoff is given by:

$$P = \Pr\{g > c\}$$

where g denotes the grade in general, and c the cutoff grade. Tables are widely available for the Standard Normal distribution, which tabulate the proportion of the Standard Normal distribution which lies *below* a given value z. For any other Normal distribution the z value is determined by taking the value of interest, subtracting the mean of the distribution, and dividing by the standard deviation. In our example, we are considering the cutoff grade, c, so that

$$z = \frac{c - \bar{g}}{s}$$

The Normal table will give us $\Phi(z)$, which is the probability of lying *below* the cutoff, so that

$$P = 1 - \Phi(z)$$

Thus, if we consider the distribution of the 'point' values, we obtain the following:

$$c = 44\,\%\,\text{Fe}$$

$$\bar{g} = 48\,\%\,\text{Fe} \qquad s = 5\,\%\,\text{Fe}$$

$$z = \frac{c - \bar{g}}{s} = \frac{44 - 48}{5} = -0{\cdot}80$$

Consulting a Standard Normal table gives $\Phi(z) = 0{\cdot}212$ so that $P = 0{\cdot}788$. That is, about 79 % of the deposit will lie above a cutoff of 44 % Fe. The second question is the average grade of this ore. For the Normal distribution, the grade above cutoff is given by:

$$\bar{g}_c = \bar{g} + \frac{s}{P}\phi(z)$$

where $\bar{g}_c$ denotes the grade above cutoff, and $\phi(z)$ is the *height* of the standard normal curve at the value z, i.e.

$$\phi(z) = \frac{1}{\sqrt{2\pi}}\exp(-z^2/2)$$

For our example, then, $\phi(z) = 0{\cdot}290$ so that:

$$\bar{g}_c = \bar{g} + \frac{s}{P}\phi(z) = 48 + \frac{5}{0{\cdot}788}0{\cdot}290 = 49{\cdot}84\,\%\,\text{Fe}$$

To summarise, 78·8 % of the ore lies above a cutoff of 44 % Fe, and this ore has an average grade of 49·8 % Fe.

Let us repeat this exercise, but now taking into account the 'selective mining unit' of 100 ft × 100 ft × 50 ft. We have:

$$c = 44\,\%\,\text{Fe}$$

$$\bar{g} = 48\,\%\,\text{Fe} \qquad \mathbf{s_v = 4{\cdot}45\,\%\,Fe}$$

$$z_v = \frac{c - \bar{g}}{s_v} = \frac{44 - 48}{4{\cdot}45} = -0{\cdot}899$$

The Standard Normal table gives $\phi(z_v) = 0{\cdot}184$ so that P is now 0·816, and the average grade above cutoff is

$$\bar{g}_c = \bar{g} + \frac{s_v}{P}\phi(z_v) = 48 + \frac{4{\cdot}45}{0{\cdot}816}0{\cdot}267 = 49{\cdot}45\,\%\,\text{Fe}$$

Although the differences in this example are not substantial, it can clearly be seen that if the selection is applied to the average grade of 100 ft by 100 ft by 50 ft blocks, then the tonnage mined (the proportion of the deposit) is *larger* and the grade of the ore is *lower* than would be expected from the original samples. If the cutoff grade chosen were above the mean grade of the deposit, the position would be reversed. This is not merely an academic observation, but is borne out by experience on many mines in existence.

Now let us turn to a much more common situation, in which the grade distribution is *log-normal*. Take as an example a lead/zinc deposit, where the 'combined metal' percentage is the economic variable. The samples are known to be log-normally distributed, and the mean and standard deviation of the 'point' samples are 12 % and 8 % combined metal respectively. The semi-variogram is spherical with a range of 15 m. The 'selective mining unit' is a block 10 by 10 by 5 m. Using Table 3.5, we can find that $F(10, 10, 5)$ when $a = 15$ and $C = 64$ is given by $0{\cdot}516 \times 64 = 33{\cdot}034$. This produces a block standard deviation of 5·56 % combined metal. The two distributions are compared in Fig. 3.13.

To calculate the proportion above cutoff and the average grade of the ore for a log-normal distribution requires an extra step in the proceedings. By definition, if a variable has a log-normal distribution, then the logarithm of that variable has a Normal distribution. It is necessary to calculate the

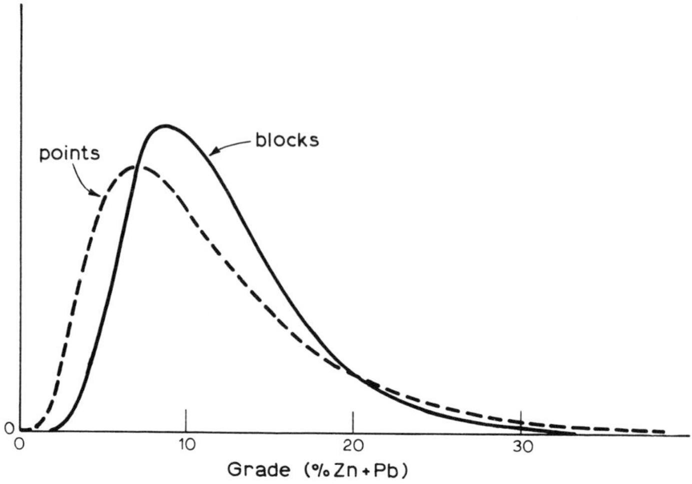

FIG. 3.13. Comparison of the distributions of combined metal percentages within a hypothetical lead/zinc deposit.

mean and standard deviation of this Normal distribution before we can perform any calculations. If we write y for the logarithm of the grade, then the mean and standard deviation of the y values are given by:

$$s_y^2 = \log_e\left(\frac{s^2}{\bar{g}^2} + 1\right)$$

$$\bar{y} = \log_e \bar{g} - 0{\cdot}5\, s_y^2$$

Once these parameters have been calculated, then the following may be evaluated:

$$z = \frac{\log_e c - \bar{y}}{s_y}$$

$$P = 1 - \Phi(z)$$

where P is again the proportion of the ore above cutoff. The average grade is found by the following process:

$$\bar{g}_c = \frac{Q}{P}\bar{g}$$

where $Q = 1 - \Phi(z - s_y)$.

In the lead/zinc example above, 4% combined metal is a feasible cutoff grade. Considering the 'point' distribution then:

$$\bar{g} = 12 \qquad s = 8$$

$$\bar{y} = 2{\cdot}30 \qquad s_y = 0{\cdot}61$$

$$z = \frac{\log_e c - \bar{y}}{s_y} = \frac{1{\cdot}39 - 2{\cdot}30}{0{\cdot}61} = -1{\cdot}508$$

$$P = 1 - \Phi(-1{\cdot}508) = \mathbf{0{\cdot}934}$$

$$Q = 1 - \Phi(-1{\cdot}508 - 0{\cdot}61) = 0{\cdot}983$$

$$\bar{g}_c = \frac{Q}{P}\bar{g} = \frac{0{\cdot}983}{0{\cdot}934} \times 12 = \mathbf{12{\cdot}62\,\%\,(Pb + Zn)}$$

The original sample data informs us that 93·4% of the deposit lies above 4% (Pb + Zn) and that this ore has an average value of 12·62% combined metal. Now, considering the distribution of 10 by 10 by 5 m blocks, the following results are found:

$$\bar{g} = 12 \qquad s_v = 5{\cdot}56$$

$$\bar{y} = 2{\cdot}39 \qquad s_y = 0{\cdot}44$$

$$z = \frac{\log_e c - \bar{y}}{s_y} = -2{\cdot}271$$

$$P = 1 - \Phi(-2{\cdot}271) = \mathbf{0{\cdot}988}$$

$$Q = 1 - \Phi(-2{\cdot}271 - 0{\cdot}44) = 0{\cdot}997$$

$$\bar{g}_c = \frac{Q}{P}\bar{g} = \frac{0{\cdot}997}{0{\cdot}988} \times 12 = \mathbf{12{\cdot}10}\,\%\,(\mathrm{Pb + Zn})$$

Once again, selection made on a mining block unit, this time of size 10 by 10 by 5 m, produces a larger tonnage and a lower grade than the small samples would suggest. Table 3.6 shows the resulting values when a *set* of possible cutoff values is chosen. Figure 3.14 shows the resulting grade/tonnage curves in the normal manner employed in mining reports. The one minor difference here is that 'proportion' above cutoff is given rather than tonnage, to keep the example general. The 'bias' introduced by considering 'point' samples rather than the true selective mining unit can clearly be seen on this graph.

Here is a very brief example with which to finish off. A low grade uranium deposit has a log-normal distribution with mean 0·30% U_3O_8 and a

TABLE 3.6. COMPARISON OF GRADE/TONNAGE CALCULATIONS FOR POINT AND BLOCK VALUES FOR COMBINED METAL (LEAD/ZINC) DEPOSIT

Cutoff	POINTS		BLOCKS	
	proportion above cutoff	*average grade*	*proportion above cutoff*	*average grade*
4	0·934	12·62	0·988	12·10
6	0·800	13·90	0·912	12·68
8	0·643	15·58	0·758	13·82
10	0·499	17·48	0·576	15·34

standard deviation of 1·05 % U_3O_8. The spherical semi-variogram has a range of 40 m, and the selective mining unit has a size of 25 by 25 by 10 m. Using Table 3.5 gives an F function of $0{\cdot}477 \times 1{\cdot}1025 = 0{\cdot}526$, producing a standard deviation for the blocks of 0·76 % U_3O_8. Table 3.7 shows the results of applying cutoffs of 0·05, 0·10, 0·15 and 0·20 % U_3O_8 to (i) the 'point' samples and (ii) the block distribution. Figures 3.15 and 3.16 illustrate these results. Notice how the skewed nature of the original

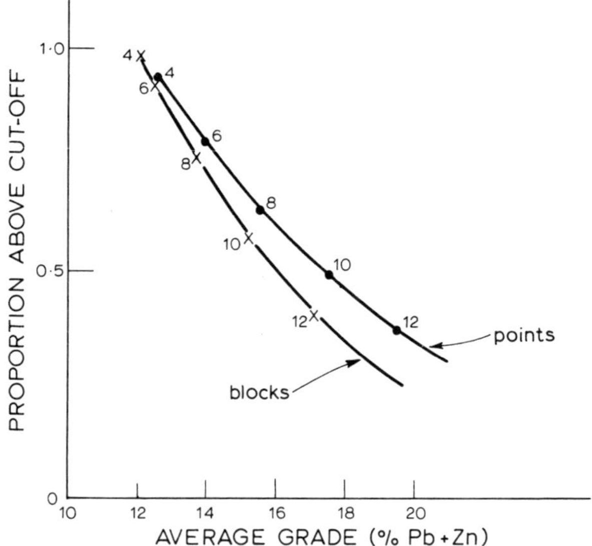

FIG. 3.14. Comparison of grade/tonnage curves in the lead/zinc deposit.

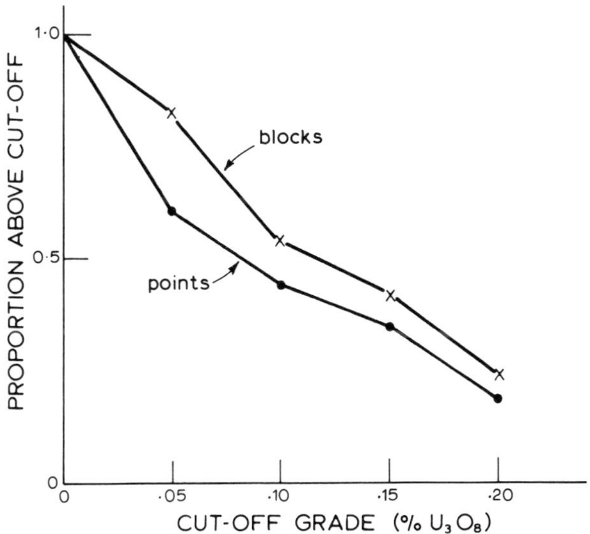

FIG. 3.15. Comparison of point and block grade/tonnage curves in a uranium deposit—cutoff grade versus proportion above cutoff.

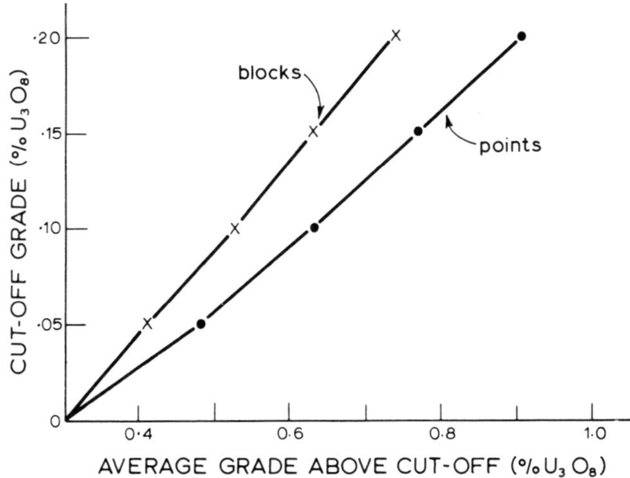

FIG. 3.16. Comparison of point and block grade/tonnage curve in a uranium deposit—cutoff grade versus average grade.

TABLE 3.7. COMPARISON OF GRADE/TONNAGE CALCULATIONS FOR POINT AND BLOCK VALUES FOR URANIUM DEPOSIT

Cutoff	POINTS		BLOCKS	
	proportion above cutoff	*average grade*	*proportion above cutoff*	*average grade*
0·05	0·622	0·47	0·712	0·41
0·10	0·452	0·62	0·527	0·53
0·15	0·355	0·75	0·414	0·64
0·20	0·291	0·88	0·337	0·75

distribution, and the relatively large size of the block combine to produce an ever widening gap between the point curve and the block curve.

CONCLUSION

In this chapter we have considered the problems introduced by the volume, size, shape etc. of both samples and selective mining units. We have seen how the 'point' semi-variogram model may be derived, even though the experimental semi-variogram may have been constructed on samples with a definite support. We have also seen how, for certain regular shapes, the distribution of values changes according to the support of the selective mining unit. The construction of 'theoretical' grade/tonnage curves can be achieved providing the following information is available:

(i) the distribution of the 'point' values;
(ii) the semi-variogram of the 'point' values;
(iii) the size and shape of the selective mining unit.

In this way more 'realistic' first estimates of the grade/tonnage calculations may be produced at a very early stage in any study.

CHAPTER 4

Estimation

So far we have used the basic concepts and assumptions of Geostatistics to build ourselves a 'model' of the structure and continuity within the deposit. We have also (in Chapter 3) seen how this can lead to the production of 'theoretical' grade/tonnage curves and the study of how mining block size can influence final production figures. It is now time we returned to our original problem of the estimation of ore reserves. The discussion in this (and the next) chapter will be confined to 'local' estimation, i.e. interest is confined to one portion of the deposit at a time. However, it should be borne in mind that the same techniques can be applied on a global scale, i.e. to the whole deposit at once. It should also be remembered that block-by-block or stope-by-stope estimates will lead inevitably to global estimates.

Let us, then, define the situation which is of interest to us. There is a point or an area or a volume of ground over which we do not know the grade (or value), but we wish to estimate it. Let us call this 'unknown' grade T, and the area (or point, or volume) of interest A. In order to produce an estimator we must have some information, usually in the form of samples. To be completely general, let us suppose n samples with values of $g_1, g_2, g_3 \ldots g_n$. This set of samples is generally denoted by S. From these samples we can form a 'linear' type of estimator—that is, a weighted average. We must restrict ourselves to this type of estimator at this stage. The estimator is denoted by T^* and is equal to:

$$T^* = w_1 g_1 + w_2 g_2 + w_3 g_3 + \cdots + w_n g_n$$

where the $w_1, w_2, w_3 \ldots w_n$ are the weights assigned to each sample. Most currently used local estimation techniques use a weighted average approach—inverse distance techniques and so on. The simplest case of all is when all of the weights are the same, and T^* is just the arithmetic mean of the sample values.

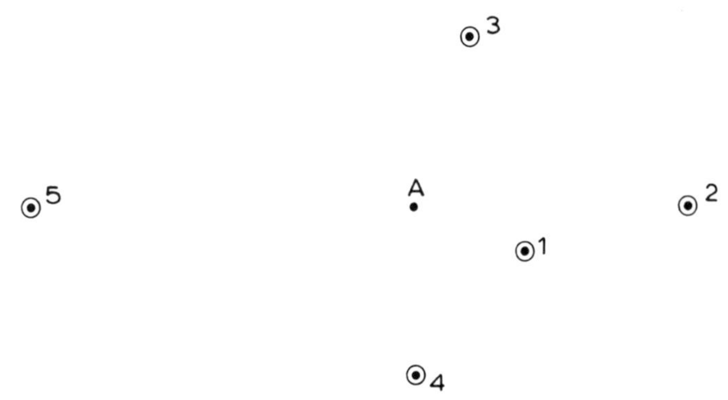

FIG. 4.1. Hypothetical sampling and estimation situation—a uranium deposit.

Now consider the setup of samples and 'unknown' which we originally discussed in the first chapter. Figure 4.1 shows the point of interest which lies at position A, and we have five 'point' samples lying around this position. The co-ordinates of these six points and the values of the samples are given in Table 4.1. The hypothetical deposit is a low-grade, large-tonnage uranium one, which is assumed to be isotropic. The semi-variogram model fitted to this deposit is a spherical one with a range of influence of 100 ft, a sill value (C) of 700 (p.p.m.)^2 and a nugget effect of 100 (p.p.m.)^2. Let us take the simplest possible estimation procedure. Take the value at the closest sample position (1) and 'extend' this to the unknown point. In doing so we incur an estimation error, ε, which will be equal to the difference between the *actual* value T and the estimated value T^*, which in this case equals g_1. That is:

$$T^* = g_1$$

$$\varepsilon = T - T^*$$

It is not too difficult to show that if there is no trend (at least locally), this estimator is unbiased. That is, if we make lots of similar estimations the average error will be zero.

$$\bar{\varepsilon} = 0$$

The 'reliability' of the estimation can be measured by looking at the *spread* of the errors. If the errors take values consistently close to zero, then the estimator is a 'good' one. If the spread of values is large, then the estimator will be unreliable. The simplest stable measure of spread (statistically) is the standard deviation. The standard deviation of an estimation error—or

TABLE 4.1. POSITIONS AND VALUES ON HYPOTHETICAL URANIUM ESTIMATION PROBLEM

Point	*Easting (ft)*	*Northing (ft)*	*Grade* U_3O_8
A	4 150	2 340	
1	4 170	2 332	400
2	4 200	2 340	380
3	4 160	2 370	450
4	4 150	2 310	280
5	4 080	2 340	320

standard error as it is referred to in ordinary statistics—will therefore measure the reliability of that estimator. No matter how many estimations we perform, we cannot calculate the standard deviation of the errors since we do not know the value of the error made. Therefore we must look at the 'theoretical' form of the variance of the estimation error, i.e. the estimation variance:

$$\varepsilon = T - T^*$$

variance of the errors $= \sigma_\varepsilon^2$

$= \text{average squared deviation from the mean error}$

$= \text{average of } (\varepsilon - \bar{\varepsilon})^2$

$= \text{average of } \varepsilon^2 \text{ since } \bar{\varepsilon} = 0$

$= \text{average of } (T - T^*)^2$

The average would be made (theoretically) over the whole deposit. That is, the same estimation situation would be repeated over the whole deposit and the variance found. This cannot be done in practice, of course, so let us look closer at the form of this variance. It is found by taking the grade at point A, subtracting the grade at point 1, squaring the result, repeating the process over all possible pairs of such points and then averaging the values. This sounds exactly like the definition of a *variogram*. In fact, it is the variogram between the two points A and (1). Given the distance between them (h) we can evaluate this estimation variance simply by reading a value from the semi-variogram model (γ) and multiplying it by 2. This is one of the reasons why it is good policy to avoid confusing the variogram and the semi-variogram. Thus:

$$\sigma_\varepsilon^2 = 2\gamma(h)$$

In the case of our particular example given in Fig. 4.1:

$$T^* = 400\,\text{p.p.m.}$$
$$h = 21{\cdot}54\,\text{ft}$$
$$\gamma(h) = 322{\cdot}7\,(\text{p.p.m.})^2$$
$$\sigma_\varepsilon^2 = 2 \times 322{\cdot}7 = 645{\cdot}4\,(\text{p.p.m.})^2$$
$$\sigma_\varepsilon = 25{\cdot}4\,\text{p.p.m.}$$

Given our knowledge about this deposit, i.e. the semi-variogram model, we can state (without too much fear of error) that the estimator used has a standard error of 25·4 p.p.m. Turning this standard error into a confidence interval, however, requires the assumption of some kind of probability distribution for the deposit. For instance if we hope that the Central Limit Theorem holds, we can say that a 95% confidence interval for T would be given by $T^* \pm 1{\cdot}96\sigma_\varepsilon$, i.e. (350 p.p.m., 450 p.p.m.). On the other hand, if we were to assume a log-normal distribution for the errors, the 95% confidence interval would be given by (354 p.p.m., 453 p.p.m.).

Now, let us complicate the procedure a little. Instead of estimating the value at the point A, in a more realistic situation (at least in mining) we would be interested in the average grade over an area or block or some mining unit. In Fig. 4.2, a 'panel' 60 ft by 30 ft has been centred on the original point A. The estimation procedure then becomes:

$$T = \text{average grade over panel}$$
$$A = \text{panel } 60 \times 30\,\text{ft}$$
$$T^* = g_1$$
$$\varepsilon = T - T^*.$$

The same arguments as previously still hold. The average error can be shown to be zero if there is no local trend. The estimation variance is still a variogram, but it is now the variogram between the grade at sample point (1) and the *average* grade over the panel A. We saw in Chapter 3 that we could cope with average grades over samples if we wanted the semi-variogram between samples of the same size, but so far we have not considered the possibility of having two different sizes to compare. The model semi-variogram supplies us with the difference in grades between two points. We could find the value of the semi-variogram between the sample point and every point within the panel A, and we could average those values.

Let us define this quantity as $\bar{\gamma}(S, A)$, read as 'gamma-bar between the sample and every point in the panel'. The 'bar' notation is the standard one for arithmetic mean. This gamma-bar term will take the place of the $\gamma(h)$ in our previous relationship. However, what we really need is the semi-variogram between the average grade of panel and the sample, *not* between all the individual points within the panel and the sample. $2\bar{\gamma}(S, A)$ would be the variance of the error made if we tried to estimate every point within the panel. To correct for this difference in emphasis we need to take into

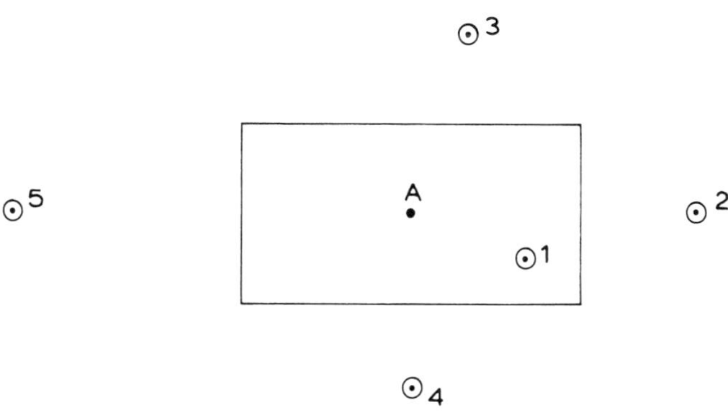

FIG. 4.2. More realistic estimation—the value of the block is required (uranium deposit).

account the variation of the grades at points within the panel. This was discussed in Chapter 3, and we evaluated it using the auxiliary function $F(l, b)$. This was the average semi-variogram between all possible pairs of points within the panel. We can rewrite this in a more general way using the gamma-bar notation. That is, $\bar{\gamma}(A, A)$ will be the average semi-variogram value between every point in the panel and every point in the panel. In the case shown in Fig. 4.2, then, when using the value at sample point (1) to estimate the average grade of the panel, the estimation variance becomes:

$$\sigma_\varepsilon^2 = 2\bar{\gamma}(S, A) - \bar{\gamma}(A, A)$$

The calculation of these gamma-bar terms will be discussed more fully later.

Now, let us complicate the mathematics still further. We actually have more than one sample available to us, so why not use them in the estimation procedure. Suppose we use the arithmetic mean of the samples as our T^*. This gives us the simplest form of the weighted average type of estimator.

That is:

$$T = \text{average grade of the panel}$$
$$A = \text{panel } 60\,\text{ft} \times 30\,\text{ft}$$
$$S = 5 \text{ point-samples at specified locations}$$
$$T^* = \tfrac{1}{5}(g_1 + g_2 + g_3 + g_4 + g_5)$$

In this case the term $\bar{\gamma}(S, A)$ is the average semi-variogram value between each point in the 'sample set' S and each point in the panel A. The term $\bar{\gamma}(A, A)$ is still the average semi-variogram between each point in the panel and each point in the panel. However, now we have yet another source of spurious variation. We only consider the average grade of the samples as the estimator, but $\bar{\gamma}(S, A)$ takes the *individual* grades into account. Thus we have also to subtract a $\bar{\gamma}(S, S)$ term from the variance, where this is the average semi-variogram value between each point in the sample set and each point in the sample set (i.e. 25 'pairs' of samples). The final version of the estimation variance then becomes:

$$\sigma_e^2 = 2\bar{\gamma}(S, A) - \bar{\gamma}(S, S) - \bar{\gamma}(A, A)$$

The arithmetic mean is often known in Geostatistics as an *extension* estimator, and the above variance is referred to as the extension variance. To distinguish this variance from the more general estimation variance for a weighted average, the subscript e is used rather than the general ε.

CALCULATION OF GAMMA-BAR TERMS

Having produced a formula for the extension variance, it only remains to explain how to calculate such terms as $\bar{\gamma}(S, A)$ in practice. For the sake of our (too) simplistic approach, we will consider for the moment only simple idealistic cases, and these only in one or two dimensions. Generalisation will be discussed later. Consider, as an example, the setup in Fig. 4.3. There is

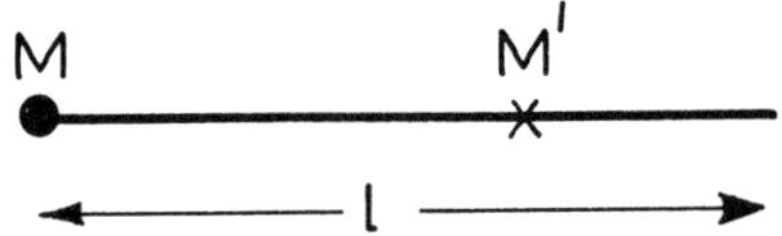

FIG. 4.3. Example of using a peripheral point to estimate the average value of the line segment.

a length of, say, drive, l m long, whose grade is unknown. We have at our disposal a single sample, perhaps at a development heading, whose value is known. In our previous notation T is the average grade over l, T^* is the grade at the sample position, A is the length and S is the single sample point. The reliability of this estimator is given by:

$$\sigma_e^2 = 2\bar{\gamma}(S, A) - \bar{\gamma}(S, S) - \bar{\gamma}(A, A)$$

$\bar{\gamma}(S, S)$ is the semi-variogram between the sample point and itself, which is zero because the sample is a 'point'. $\bar{\gamma}(A, A)$ is none other than the $F(l)$ function encountered in Chapter 3. Our problem arises with $\bar{\gamma}(S, A)$ which has been defined as the average semi-variogram between the sample point and every point in the line. That is, we take M as a fixed point (the sample) and M$'$ can be anywhere on the line. We take all such pairs that are possible, calculate the value of the semi-variogram for each pair, sum these (using an integration), and average this sum. Because the 'sum' is being performed over a continuous length, we cannot divide it by the 'number of points' in the sum. Instead we divide by the length of the line itself, l. This produces another auxiliary function which is called $\chi(l)$ and deals with the specific case of points on the end of lines. Thus our extension variance becomes:

$$\sigma_e^2 = 2\chi(l) - F(l)$$

It remains only to determine the function $\chi(l)$ for the particular model in use and the standard error is immediately available. The one-dimensional auxiliary functions are given below for the three 'common' models. Semi-variograms comprising more than one component model are easily handled. The auxiliary function for each component is evaluated and then the component auxiliary functions added together.

Auxiliary functions

Linear model for the semi-variogram;

$$\gamma(h) = ph$$

$$\chi(l) = p\frac{l}{2}$$

$$F(l) = p\frac{l}{3}$$

Exponential model for the semi-variogram;

$$\gamma(h) = C[1 - \exp(-h/a)]$$

$$\chi(l) = C\left\{1 - \frac{a}{l}[1 - \exp(-l/a)]\right\}$$

$$F(l) = C\left\{1 - \frac{a}{l} + \frac{a^2}{l^2}[1 - \exp(-l/a)]\right\}$$

Spherical model for the semi-variogram;

$$\gamma(h) = C\left(\frac{3}{2}\frac{h}{a} - \frac{1}{2}\frac{h^3}{a^3}\right) \quad \text{when } h \leq a$$

$$= C \quad \text{when } h \geq a$$

$$\chi(l) = \frac{C}{8}\frac{l}{a}\left(6 - \frac{l^2}{a^2}\right) \quad \text{when } l \leq a$$

$$= \frac{C}{8}\left(8 - 3\frac{a}{l}\right) \quad \text{when } l \geq a$$

$$F(l) = \frac{C}{20}\frac{l}{a}\left(10 - \frac{l^2}{a^2}\right) \quad \text{when } l \leq a$$

$$= \frac{C}{20}\left(20 - 15\frac{a}{l} + 4\frac{a^2}{l^2}\right) \quad \text{when } l \geq a$$

Thus in our example above, if we have a linear semi-variogram, the extension variance for the setup in Fig. 4.3 becomes:

$$\sigma_e^2 = 2p\frac{l}{2} - p\frac{l}{3} = \frac{2}{3}pl$$

For any specific problem, we need specify only the length of the line l and the slope of the semi-variogram, p.

Let us now consider a slightly more interesting example, such as that shown in Fig. 4.4. Here the point sample is in the *middle* of the line, but otherwise the situation remains the same. In:

$$\sigma_e^2 = 2\bar{\gamma}(S, A) - \bar{\gamma}(S, S) - \bar{\gamma}(A, A)$$

only the first term $\bar{\gamma}(S, A)$ has changed. Rather than invent a new auxiliary function, or have to do the integration all over again, we can use the existing

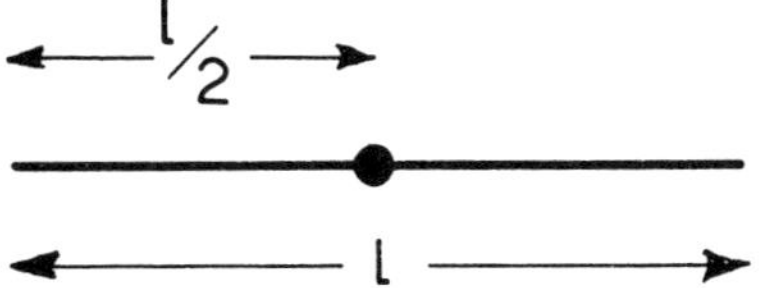

FIG. 4.4. Example of using a central point to estimate the average value of the line segment.

$\chi(l)$ function to produce the required term. The term we require is as follows:

$\bar{\gamma}(S, A)$ = the average semi-variogram value between the sample point and every point along the line.
= (sum of all the semi-variogram values between the sample point and every point along the line)/l.
= (sum of all the semi-variogram values between the sample point and every point in the left hand half of the line + sum of all the values between the sample and the right hand half of the line)/l.

Figure 4.5 illustrates the 'splitting' of the line so as to put the sample point at the *end* of two shorter lines. Now, $\chi(l/2)$ would give us the average of all the semi-variogram values between M (the sample point) and the M′ on the left hand half of the line. Returning to the definition of the χ function, it can easily be seen that the *sum* of all the semi-variogram values between M and M′ will be the *average* multiplied by the length of line under consideration.

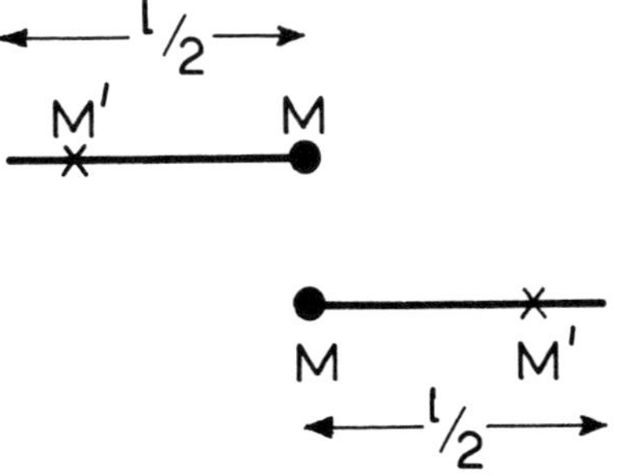

FIG. 4.5. Simplifying the central point problem to allow the use of auxiliary functions.

Thus:

$$\bar{\gamma}(S, A) = \frac{1}{l}\left[\frac{l}{2}\chi(l/2) + \frac{l}{2}\chi(l/2)\right] = \chi(l/2)$$

so that

$$\sigma_e^2 = 2\chi(l/2) - F(l)$$

In a particular case the user may substitute his own model for the semi-variogram, and hence the appropriate auxiliary functions. Before moving on let us compare this result with the previous situation, where the sample lay at the *end* of the line. In the former case the extension variance was:

$$\sigma_e^2 = 2\chi(l) - F(l)$$

By definition $\chi(l)$ must be greater than (or at least equal to) $\chi(l/2)$. The conclusion? If you can only take one sample, it is better to take it in the middle of what you are trying to estimate. It is reassuring to find that so-called common sense has a sound mathematical background.

FIG. 4.6. Generalisation of the 'central' point problem.

Using the same sort of logic on Fig. 4.6, you should be able to deduce that:

$$\bar{\gamma}(S, A) = \frac{1}{l}[b\chi(b) + (l - b)\chi(l - b)]$$

so that

$$\sigma_e^2 = \frac{2}{l}[b\chi(b) + (l - b)\chi(l - b)] - F(l)$$

Figure 4.7 at first sight seems to be a different kettle of fish. However, let us follow the same procedure and see where it leads.

$\bar{\gamma}(S, A)$ = average semi-variogram value between the point and all the points in the length l.
= (sum of the semi-variogram values between the point and all the points in the length l)/l.

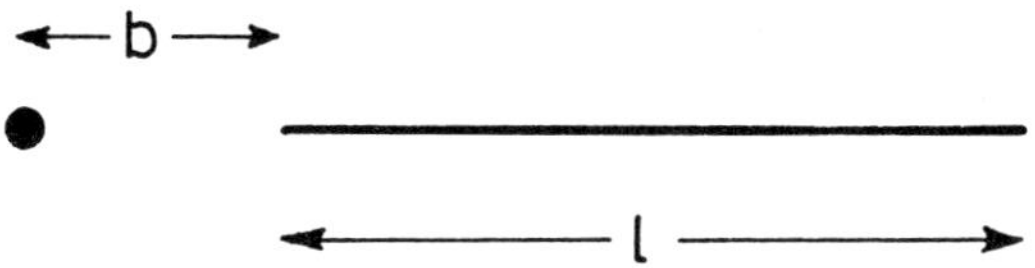

FIG. 4.7. Extrapolation of the peripheral point problem.

The point lies on the end of a 'line' of length $l + b$. The expression $(l + b)\chi(l + b)$ would give us the sum of all the semi-variogram values between the sample and the length $l + b$. However we do not require the points corresponding to M′ within the length b, so we may subtract those in the form $b\chi(b)$. That is,

$$\bar{\gamma}(S, A) = \frac{1}{l}[(l + b)\chi(l + b) - b\chi(b)]$$

so that

$$\sigma_e^2 = \frac{2}{l}[(l + b)\chi(l + b) - b\chi(b)] - F(l)$$

For the linear model, for example, this would be:

$$\sigma_e^2 = \frac{2}{l}\left[(l + b)p\frac{(l + b)}{2} - bp\frac{b}{2}\right] - p\frac{l}{3}$$

$$= \frac{p}{l}(l^2 + 2lb + b^2 - b^2) - p\frac{l}{3}$$

$$= p(l + 2b) - p\frac{l}{3} = \frac{2}{3}pl + 2pb$$

This is obviously larger than the expression when the point was on the end of the line, as would be expected.

One last example before we abandon one-dimensional examples: Fig. 4.8 shows the 'same' line, which now contains three samples. We

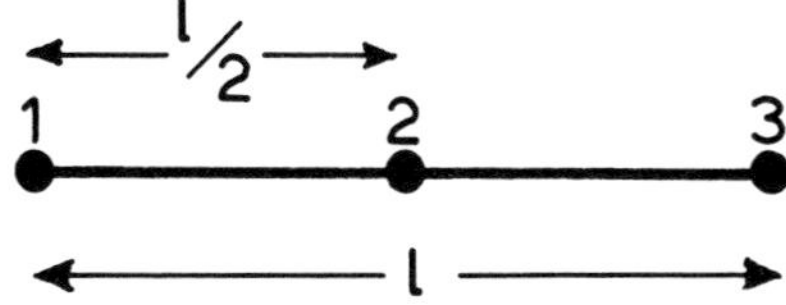

FIG. 4.8. More complex problem when three samples are available to estimate the line segment.

shall use the arithmetic mean of the three grades to estimate the length, i.e. $T^* = \frac{1}{3}(g_1 + g_2 + g_3)$. Then our extension variance is

$$\sigma_e^2 = 2\bar{\gamma}(S, A) - \bar{\gamma}(S, S) - \bar{\gamma}(A, A)$$

where S is now a set of three points. $\bar{\gamma}(A, A)$ remains unchanged, equal to $F(l)$ since we have not changed the length to be estimated at all. However, $\bar{\gamma}(S, A)$ is now the average semi-variogram value between each of the three points and the line, so that

$$\bar{\gamma}(S, A) = \tfrac{1}{3}[\bar{\gamma}(S_1, A) + \bar{\gamma}(S_2, A) + \bar{\gamma}(S_3, A)]$$

where S_1 represents sample 1 and so on. Now $\bar{\gamma}(S_1, A)$ is simply $\chi(l)$, as is $\bar{\gamma}(S_3, A)$. The term $\bar{\gamma}(S_2, A)$ is the same situation as that in Fig. 4.4, so this equals $\chi(l/2)$. Thus,

$$\bar{\gamma}(S, A) = \tfrac{1}{3}(2\chi(l) + \chi(l/2))$$

The middle term of the variance $\bar{\gamma}(S, S)$ requires us to take each point in the sample set with each point in the sample set. Since there are three points in the set, there are nine such pairs of points:

$$\begin{aligned}\bar{\gamma}(S, S) = \tfrac{1}{9}[&\gamma(S_1, S_1) + \gamma(S_1, S_2) + \gamma(S_1, S_3)\\ &\gamma(S_2, S_1) + \gamma(S_2, S_2) + \gamma(S_2, S_3)\\ &\gamma(S_3, S_1) + \gamma(S_3, S_2) + \gamma(S_3, S_3)]\end{aligned}$$

Each of the individual terms is simply the semi-variogram between a pair of points. Three of the terms, $\gamma(S_1, S_1)$, $\gamma(S_2, S_2)$ and $\gamma(S_3, S_3)$ are automatically zero since the samples are points. The terms $\gamma(S_1, S_2)$, $\gamma(S_2, S_1)$, $\gamma(S_2, S_3)$ and $\gamma(S_3, S_2)$ are all equal to $\gamma(l/2)$, whilst $\gamma(S_1, S_3)$ and $\gamma(S_3, S_1)$ are equal to $\gamma(l)$. Thus:

$$\bar{\gamma}(S, S) = \tfrac{1}{9}[4\gamma(l/2) + 2\gamma(l)]$$

so that

$$\sigma_e^2 = \tfrac{2}{3}[2\chi(l) + \chi(l/2)] - \tfrac{1}{9}[4\gamma(l/2) + 2\gamma(l)] - F(l)$$

For our linear example, this reduces to:

$$\sigma_e^2 = \frac{2}{3}\left(2p\frac{l}{2} + p\frac{l/2}{2}\right) - \frac{1}{9}\left(4p\frac{l}{2} + 2pl\right) - p\frac{l}{3}$$

$$= \frac{2}{3}pl + p\frac{l}{6} - \frac{2}{9}pl - \frac{2}{9}pl - \frac{1}{3}pl$$

$$= \frac{pl}{18}$$

This is considerably less than the other evaluations, as seems only sensible with three times as many samples.

TWO-DIMENSIONAL EXAMPLES

In two dimensions we have again a set of auxiliary functions which are mostly generalisations of the one-dimensional ones. These are shown in Fig. 4.9. $\gamma(l;b)$ is the two-dimensional analogue of $\gamma(l)$—sometimes read as '$\gamma(l)$ stretched through length b'. It is the average semi-variogram value between all points on one length, b, and all points on another parallel to it and of the same length. This function is useful for parallel boreholes, channel samples, drives, raises and so on. The generalisation of $\chi(l)$ becomes $\chi(l;b)$ and is the average semi-variogram between a length b (drive, raise, core etc.) and an adjacent panel l by b. The function $F(l, b)$ has been introduced in Chapter 3 for such terms as $\bar{\gamma}(A, A)$ for rectangular panels. We also introduce a 'new' function $H(l, b)$ which does duty for two rather different situations. It represents the average semi-variogram value between a panel and a point on one corner of it; it *also* represents the average semi-variogram value between two lengths l and b at right angles to one another. This is simply a mathematical accident.

A few examples will be given here to show how to 'manoeuvre' situations into the required form. Figure 4.10 shows a stoping panel in a cassiterite vein, 100 ft by 125 ft, with a development drive passing through it *parallel* to the sides. The drive is 25 ft from the 'bottom' of the panel. The drive is so closely sampled that we can assume we know its average grade. We wish to use that average grade as an estimate of the average grade of the panel, so that:

T = average grade of the panel

A = panel 100 ft by 125 ft

T^* = average grade of the interior drive

S = 125 ft long drive

$$\sigma_e^2 = 2\bar{\gamma}(S, A) - \bar{\gamma}(S, S) - \bar{\gamma}(A, A)$$

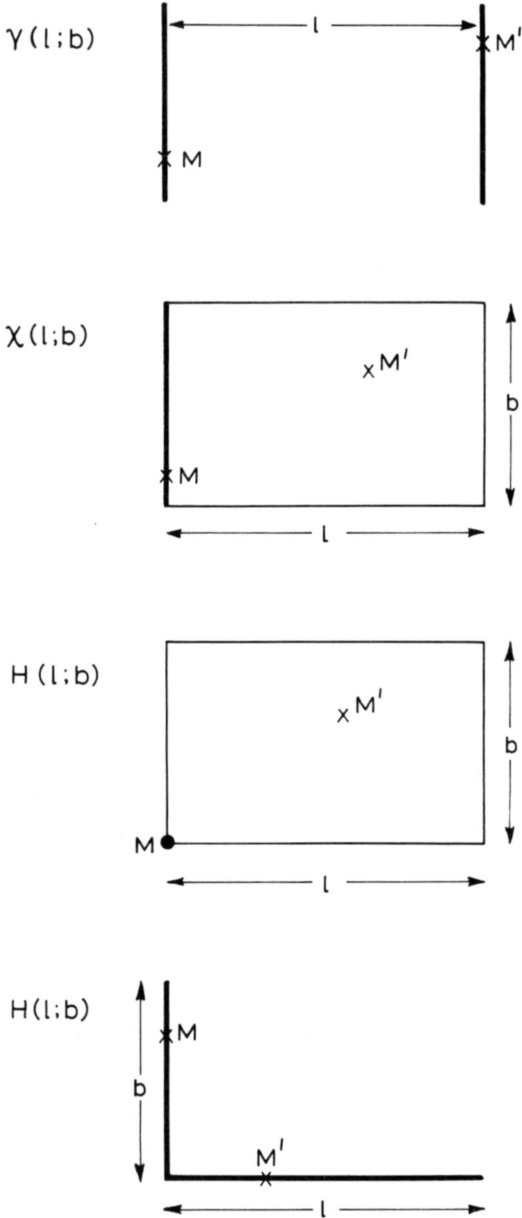

FIG. 4.9. The two-dimensional auxiliary functions.

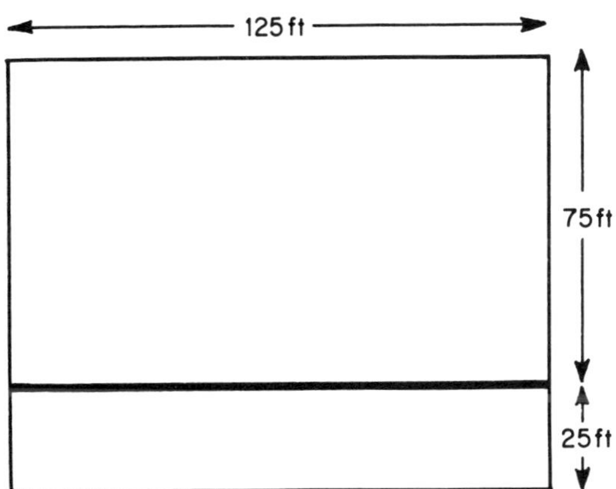

FIG. 4.10. Estimation of the panel average from the drive average.

$\bar{\gamma}(A, A)$ is the average semi-variogram value between every point in A and every point in A, which is the definition of the two-dimensional function $F(l, b)$. Thus if we have a model for the semi-variogram we can evaluate this. Let us suppose that in the above example the semi-variogram is a spherical model with a range of influence of 80 ft and a sill of 450 (lb/ton)^2. The table given in Chapter 3 for the $F(l, b)$ function (Table 3.4) is for the 'standardised' spherical model with $a = 1$ and $C = 1$. We require $F(125, 100)$ for a model in which $a = 80$ and $C = 450$. To convert to $a = 1$ we divide the measurements l and b by the range a (80 ft). Thus we require $F(125/80, 100/80) = F(1{\cdot}54, 1{\cdot}20)$ for $a = 1$ and $C = 450$. From the table (interpolating linearly) we find that $F(1{\cdot}54, 1{\cdot}20) = 0{\cdot}7807$ when $C = 1$. Therefore, if $C = 450$, our F value must be $0{\cdot}7807 \times 450 = 351{\cdot}3$ (lb/ton)^2. This finally is $\bar{\gamma}(A, A)$.

Now let us consider the term $\bar{\gamma}(S, S)$. Our sample is a 'line' of length 125 ft, and the $\bar{\gamma}(S, S)$ is the average semi-variogram value between every point in the line and every point in the line, i.e. the one-dimensional function $F(l)$. For a spherical semi-variogram model, where the length is greater than the range of influence, we know that:

$$F(l) = \frac{C}{20}\left(20 - 15\frac{a}{l} + 4\frac{a^2}{l^2}\right)$$

Substituting the values $l = 125$ ft, $a = 80$ ft and $C = 450$ (lb/ton)^2, gives $F(125) = 270{\cdot}9$ (lb/ton)^2, and this is the value used for $\bar{\gamma}(S, S)$.

Finally, we must turn to $\bar{\gamma}(S, A)$ which presents a problem, since the auxiliary function $\chi(l;b)$ is only for situations where the 'line' is on one side of the panel. We must employ the same sort of manoeuvre as before:

$\bar{\gamma}(S, A)$ = average semi-variogram value between the line and a panel 125 ft × 100 ft.

= (sum of the semi-variogram values between the line and the panel)/(100 × 125).

= (sum of the semi-variogram values between the line and the points in the 'upper' panel + sum of the semi-variogram values between the line and the points in the lower panel)/(100 × 125).

However, the average semi-variogram value between the line and the 'upper' panel (125 × 75 ft) is given by $\chi(75; 125)$, so that the sum of the semi-variogram values between the line and every point in the upper panel will be $75 \times 125 \times \chi(75; 125)$. Similarly, the sum of the semi-variogram values between the line and every point in the 'lower' panel (125 ft × 25 ft) will be given by $25 \times 125 \times \chi(25; 125)$. This gives:

$$\bar{\gamma}(S, A) = [75 \times 125 \times \chi(75; 125) + 25 \times 125 \times \chi(25; 125)]/(100 \times 125)$$

$$= \tfrac{3}{4}\chi(75; 125) + \tfrac{1}{4}\chi(25; 125)$$

Table 4.2 shows the values of the $\chi(l;b)$ function for the 'standardised' spherical model with $a = 1$ and $C = 1$. This table is used in the same way as the $F(l, b)$ table. It should be noted that the order of the arguments is important. The value of $\chi(25; 125)$ is obviously different from the value of $\chi(125; 25)$. Standardising the measurements of the panel in the same way as before, we require the values of $\chi(0{\cdot}94; 1{\cdot}54)$ and $\chi(0{\cdot}31; 1{\cdot}54)$ from the table. Interpolating linearly gives the values 0·8196 and 0·6538 respectively. Multiplying by the sill of 450 (lb/ton)^2 gives 368·8 (lb/ton)^2 and 294·2 (lb/ton)^2 respectively. Putting these together in the expression for $\bar{\gamma}(S, A)$ gives a value of 350·2 (lb/ton)^2.

Having evaluated all of the individual terms, we can now calculate the extension variance for the setup in Fig. 4.10, so that:

$$\sigma_e^2 = (2 \times 350{\cdot}2) - 351{\cdot}3 - 270{\cdot}9 = 78{\cdot}2\,(\text{lb/ton})^2$$

This gives an estimation standard deviation or 'standard error' of 8·8 lb/ton for the estimation of the panel grade. If we are willing to assume Normality

L	B = .1	.2	.3	.4	.5	.6	.7	.8	.9	1.0
.10	.098	.136	.178	.222	.266	.309	.350	.390	.428	.464
.20	.164	.194	.229	.268	.307	.346	.385	.422	.458	.491
.30	.233	.257	.288	.321	.356	.392	.427	.462	.495	.526
.40	.302	.322	.348	.378	.409	.441	.474	.505	.535	.564
.50	.368	.385	.408	.434	.462	.492	.521	.550	.577	.603
.60	.430	.445	.466	.489	.515	.541	.568	.594	.619	.642
.70	.488	.502	.520	.541	.564	.588	.612	.636	.658	.680
.80	.542	.554	.570	.589	.610	.631	.653	.674	.695	.714
.90	.589	.600	.614	.632	.650	.670	.689	.708	.727	.744
1.00	.629	.639	.653	.668	.685	.703	.720	.737	.754	.769
1.20	.691	.699	.711	.723	.737	.752	.767	.781	.795	.808
1.40	.735	.742	.752	.763	.775	.788	.800	.812	.824	.835
1.60	.768	.775	.783	.793	.803	.814	.825	.836	.846	.856
1.80	.794	.800	.807	.816	.825	.835	.845	.854	.863	.872
2.00	.815	.820	.826	.834	.842	.851	.860	.869	.877	.885
2.50	.852	.856	.861	.867	.874	.881	.888	.895	.902	.908
3.00	.876	.880	.884	.889	.895	.901	.907	.912	.918	.923
3.50	.894	.897	.901	.905	.910	.915	.920	.925	.930	.934
4.00	.907	.910	.913	.917	.921	.926	.930	.934	.938	.942
5.00	.926	.928	.931	.934	.937	.941	.944	.947	.951	.954

L	B = 1.2	1.4	1.6	1.8	2.0	2.5	3.0	3.5	4.0	5.0
.10	.526	.577	.619	.653	.683	.738	.777	.807	.829	.861
.20	.550	.598	.638	.671	.698	.751	.788	.816	.837	.868
.30	.580	.625	.662	.693	.719	.768	.803	.828	.848	.877
.40	.614	.655	.689	.718	.741	.787	.819	.842	.861	.887
.50	.649	.687	.718	.743	.765	.806	.835	.857	.873	.897
.60	.684	.718	.746	.769	.788	.825	.852	.871	.886	.907
.70	.717	.747	.772	.793	.811	.844	.867	.885	.898	.917
.80	.747	.774	.797	.815	.831	.861	.881	.897	.909	.926
.90	.774	.798	.818	.835	.849	.875	.894	.908	.919	.934
1.00	.796	.818	.836	.851	.864	.888	.905	.917	.927	.941
1.20	.830	.848	.864	.876	.886	.906	.920	.931	.939	.950
1.40	.854	.870	.883	.894	.903	.920	.932	.941	.948	.958
1.60	.873	.886	.898	.907	.915	.930	.940	.948	.954	.963
1.80	.887	.899	.909	.917	.924	.938	.947	.954	.959	.967
2.00	.898	.909	.918	.926	.932	.944	.952	.959	.963	.970
2.50	.918	.927	.934	.940	.946	.955	.962	.967	.971	.976
3.00	.932	.939	.945	.950	.955	.963	.968	.972	.976	.980
3.50	.942	.948	.953	.957	.961	.968	.973	.976	.979	.983
4.00	.949	.955	.959	.963	.966	.972	.976	.979	.982	.985
5.00	.959	.964	.967	.970	.973	.978	.981	.983	.985	.988

for the errors (*sic*) we could be 95% certain that the 'true' grade of the panel lay between $T^* \pm 2\sigma_e$ equal to $T^* \pm 16{\cdot}6$ lb/ton. This standard error would be correct for any panel having the same sampling setup, anywhere within this deposit, since the estimation variance depends *not* on the actual grades involved but on the geometric positioning of the sample and the panel.

As a second example, consider a disseminated nickel deposit in the late stages of development. On a particular underground level, we have a stoping block 40 m by 30 m to be estimated. If we consider only the one level we can treat the problem as a two-dimensional one. Suppose this 'panel' has been developed along two sides, and the information available consists of (i) the average grade along the 40 m drive, g_1 and (ii) the average grade along the 30 m drive, g_2. Suppose we use the average of these two grades to estimate the value inside the panel. Then:

$$\begin{aligned} T &= \text{average grade of the panel} \\ A &= \text{panel } 30\,\text{m} \times 40\,\text{m} \\ T^* &= \tfrac{1}{2}(g_1 + g_2) \\ S &= 40\,\text{m drive, } 30\,\text{m drive} \end{aligned}$$

For this particular deposit we have a spherical semi-variogram with a range of influence of 60 m, a sill of $0{\cdot}75\,(\%)^2$ and a nugget effect of $0{\cdot}10\,(\%)^2$. This implies a standard deviation for the 'point' sample values of $0{\cdot}92\%$. The extension variance, as always, is given by:

$$\sigma_e^2 = 2\bar{\gamma}(S, A) - \bar{\gamma}(S, S) - \bar{\gamma}(A, A)$$

$\bar{\gamma}(A, A)$ is, as before, the two-dimensional function $F(l, b)$, but now we have two components to the evaluation of this term. We can calculate the $F(40, 30)$ for the spherical part of the model, with $a = 60$ m and $C = 0{\cdot}75\,(\%)^2$. This turns out to be $0{\cdot}4336 \times 0{\cdot}75 = 0{\cdot}325\,(\%)^2$. To this we must add the nugget effect of $0{\cdot}10\,(\%)^2$, so that $\bar{\gamma}(A, A) = 0{\cdot}425\,(\%)^2$.

The average semi-variogram value between each sample and each sample now contains *four* terms which have to be averaged, i.e.:

$$\begin{aligned} \bar{\gamma}(S, S) = \tfrac{1}{4}[&\bar{\gamma}(\text{drive}_1, \text{drive}_1) + \bar{\gamma}(\text{drive}_1, \text{drive}_2) \\ &+ \bar{\gamma}(\text{drive}_2, \text{drive}_1) + \bar{\gamma}(\text{drive}_2, \text{drive}_2)] \end{aligned}$$

It is generally easier to look at each term separately and then to combine them to produce the final answer. The first term $\bar{\gamma}(\text{drive}_1, \text{drive}_1)$ is the average semi-variogram between all points within a length of 40 m. That is,

L \ B	.1	.2	.3	.4	.5	.6	.7	.8	.9	1.0
.10	.114	.177	.243	.310	.374	.436	.494	.546	.593	.633
.20	.177	.227	.285	.346	.406	.464	.518	.568	.613	.651
.30	.243	.285	.336	.390	.445	.499	.550	.597	.639	.674
.40	.310	.346	.390	.439	.489	.539	.586	.629	.668	.701
.50	.374	.406	.445	.489	.535	.580	.623	.663	.698	.728
.60	.436	.464	.499	.539	.580	.621	.660	.697	.728	.755
.70	.494	.518	.550	.586	.623	.660	.696	.729	.757	.781
.80	.546	.568	.597	.629	.663	.697	.729	.758	.783	.805
.90	.593	.613	.639	.668	.698	.728	.757	.783	.806	.826
1.00	.633	.651	.674	.701	.728	.755	.781	.805	.826	.843
1.20	.694	.709	.729	.751	.774	.796	.818	.837	.855	.869
1.40	.738	.751	.767	.786	.806	.825	.844	.861	.875	.888
1.60	.771	.782	.797	.813	.830	.847	.863	.878	.891	.902
1.80	.796	.806	.819	.834	.849	.864	.879	.892	.903	.913
2.00	.817	.826	.837	.850	.864	.878	.891	.902	.913	.921
2.50	.853	.860	.870	.880	.891	.902	.913	.922	.930	.937
3.00	.878	.884	.891	.900	.909	.918	.927	.935	.942	.948
3.50	.895	.900	.907	.914	.922	.930	.938	.944	.950	.955
4.00	.908	.913	.919	.925	.932	.939	.945	.951	.956	.961
5.00	.927	.930	.935	.940	.946	.951	.956	.961	.965	.969

L \ B	1.2	1.4	1.6	1.8	2.0	2.5	3.0	3.5	4.0	5.0
.10	.694	.738	.771	.796	.817	.853	.878	.895	.908	.927
.20	.709	.751	.782	.806	.826	.860	.884	.900	.913	.930
.30	.729	.767	.797	.819	.837	.870	.891	.907	.919	.935
.40	.751	.786	.813	.834	.850	.880	.900	.914	.925	.940
.50	.774	.806	.830	.849	.864	.891	.909	.922	.932	.946
.60	.796	.825	.847	.864	.878	.902	.918	.930	.939	.951
.70	.818	.844	.863	.879	.891	.913	.927	.938	.945	.956
.80	.837	.861	.878	.892	.902	.922	.935	.944	.951	.961
.90	.855	.875	.891	.903	.913	.930	.942	.950	.956	.965
1.00	.869	.888	.902	.913	.921	.937	.948	.955	.961	.969
1.20	.891	.907	.918	.927	.935	.948	.956	.963	.967	.974
1.40	.907	.920	.930	.938	.944	.955	.963	.968	.972	.978
1.60	.918	.930	.939	.945	.951	.961	.967	.972	.975	.980
1.80	.927	.938	.945	.952	.956	.965	.971	.975	.978	.983
2.00	.935	.944	.951	.956	.961	.969	.974	.978	.980	.984
2.50	.948	.955	.961	.965	.969	.975	.979	.982	.984	.987
3.00	.956	.963	.967	.971	.974	.979	.983	.985	.987	.990
3.50	.963	.968	.972	.975	.978	.982	.985	.987	.989	.991
4.00	.967	.972	.975	.978	.980	.984	.987	.989	.990	.992
5.00	.974	.978	.980	.983	.984	.987	.990	.991	.992	.994

$F(40)$. For a spherical model, when the length of the 'line' is shorter than the range of influence, $F(l)$ is given by:

$$F(l) = \frac{C}{20}\frac{l}{a}\left(10 - \frac{l^2}{a^2}\right)$$

Substituting $l = 40$ and $a = 60$, $C = 0{\cdot}75\,(\%)^2$ gives a value of $0{\cdot}239\,(\%)^2$. Adding on the nugget effect produces $\bar{\gamma}(\text{drive}_1, \text{drive}_1) = 0{\cdot}339\,(\%)^2$. Similarly, $\bar{\gamma}(\text{drive}_2, \text{drive}_2) = 0{\cdot}283\,(\%)^2$. The two terms $\bar{\gamma}(\text{drive}_1, \text{drive}_2)$ and $\bar{\gamma}(\text{drive}_2, \text{drive}_1)$ are identical and have been defined as the auxiliary function $H(l, b)$. Using Table 4.3 for the $H(l, b)$ function in the same way as the other tables and adding in the nugget effect gives:

$$\bar{\gamma}(\text{drive}_1, \text{drive}_2) = 0{\cdot}6086 \times 0{\cdot}75 + 0{\cdot}10\,(\%)^2 = 0{\cdot}556\,(\%)^2$$

Putting all the individual terms together, we find that $\bar{\gamma}(S, S) = 0{\cdot}428\,(\%)^2$.

Finally, to calculate the estimation variance, we also require the term $\bar{\gamma}(S, A)$. This is the average semi-variogram value between each sample and the panel, so that:

$$\begin{aligned}\bar{\gamma}(S, A) &= \tfrac{1}{2}[\bar{\gamma}(\text{drive}_1, A) + \bar{\gamma}(\text{drive}_2, A)] \\ &= \tfrac{1}{2}[\chi(30;40) + \chi(40;30)] \\ &= \tfrac{1}{2}[(0{\cdot}5112 \times 0{\cdot}75 + 0{\cdot}10) + (0{\cdot}5447 \times 0{\cdot}75 + 0{\cdot}10)] \\ &= 0{\cdot}497\,(\%)^2\end{aligned}$$

The extension variance for the problem illustrated in Fig. 4.11 is thus:

$$\sigma_e^2 = (2 \times 0{\cdot}497) - 0{\cdot}428 - 0{\cdot}425 = 0{\cdot}141\,(\%)^2$$

giving a standard error for the prediction of the panel average of $0{\cdot}376\,\%$ nickel.

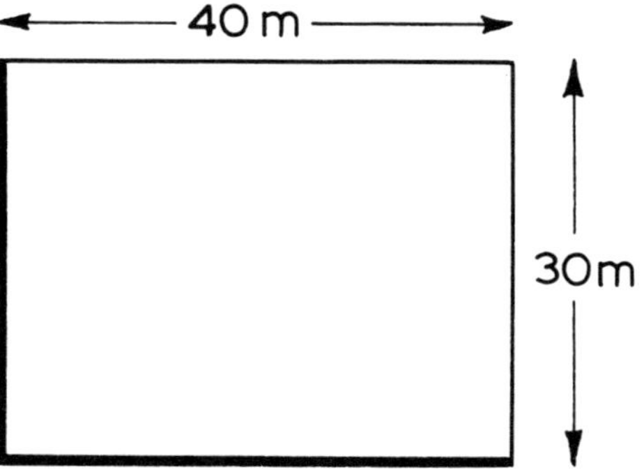

FIG. 4.11. Estimation of the panel average from two drive averages.

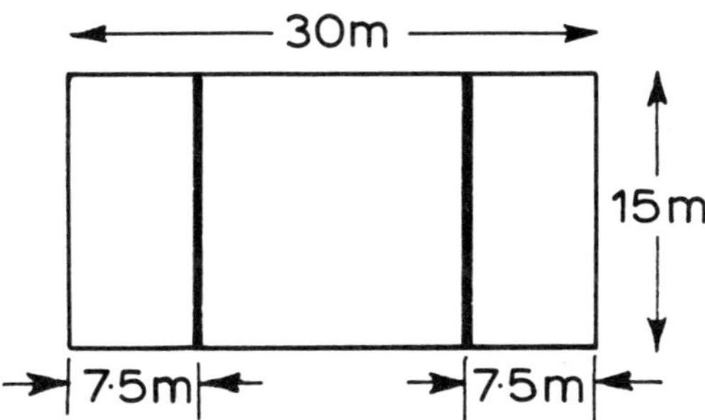

FIG. 4.12. Estimation of the panel average from two raise averages.

A third example is shown in Fig. 4.12. This time the metal is zinc, the semi-variogram is spherical with a range of 20 m and a sill of 49 (%)^2. The value to be estimated is the average over the 30 m by 15 m panel, and the information available is the average grade of each of the two raises through the 'stope panel'. In the idealised situation chosen, the raises are both 7·5 m from the edges of the panel. Very briefly, the extension variance is produced as follows:

$$T = \text{average grade of the panel}$$
$$A = \text{panel } 30\,\text{m} \times 15\,\text{m}$$
$$T^* = \tfrac{1}{2}(g_1 + g_2)$$
$$S = \text{raise}_1, \text{raise}_2$$
$$\sigma_e^2 = 2\bar{\gamma}(S, A) - \bar{\gamma}(S, S) - \bar{\gamma}(A, A)$$

The term $\bar{\gamma}(A, A) = F(30, 15)$ when $a = 20$ and $C = 49\,(\%)^2$, which is $0{\cdot}7021 \times 49 = 34{\cdot}4\,(\%)^2$.

$$\begin{aligned}\bar{\gamma}(S, S) &= \tfrac{1}{4}[\bar{\gamma}(\text{raise}_1, \text{raise}_1) + \bar{\gamma}(\text{raise}_1, \text{raise}_2) \\ &\qquad + \bar{\gamma}(\text{raise}_2, \text{raise}_1) + \bar{\gamma}(\text{raise}_2, \text{raise}_2)] \\ &= \tfrac{1}{4}[F(15) + \gamma(15;15) + \gamma(15;15) + F(15)] \\ &= \tfrac{1}{4}(17{\cdot}34 + 46{\cdot}02 + 46{\cdot}02 + 17{\cdot}34) \\ &= 31{\cdot}7\,(\%)^2\end{aligned}$$

The $\gamma(l;b)$ function is given for the 'standardised' spherical model in Table 4.4. The last term to be evaluated is:

$$\bar{\gamma}(S, A) = \tfrac{1}{2}[\bar{\gamma}(\text{raise}_1, A) + \bar{\gamma}(\text{raise}_2, A)]$$

TABLE 4.4. AUXILIARY FUNCTION $\gamma(L, B)$ FOR SPHERICAL MODEL WITH RANGE 1·0 AND SILL 1·0

L \ B	.1	.2	.3	.4	.5	.6	.7	.8	.9	1.0
.05	.094	.132	.175	.219	.263	.306	.348	.388	.426	.461
.10	.161	.188	.223	.261	.300	.340	.379	.416	.452	.486
.15	.231	.252	.280	.312	.347	.383	.419	.453	.486	.518
.20	.302	.318	.341	.369	.400	.432	.464	.495	.526	.555
.25	.372	.385	.404	.428	.455	.483	.512	.541	.568	.594
.30	.440	.451	.467	.488	.511	.536	.562	.588	.613	.636
.35	.507	.516	.529	.547	.568	.590	.612	.635	.657	.678
.40	.571	.578	.590	.605	.623	.642	.662	.683	.702	.721
.45	.632	.638	.648	.661	.677	.693	.711	.729	.746	.762
.50	.689	.695	.703	.715	.728	.742	.758	.773	.787	.801
.55	.743	.748	.755	.765	.776	.789	.802	.814	.827	.838
.60	.793	.797	.803	.811	.821	.831	.842	.853	.863	.872
.65	.839	.842	.847	.854	.862	.870	.879	.888	.896	.903
.70	.879	.882	.886	.892	.898	.905	.912	.919	.925	.930
.75	.915	.917	.920	.925	.930	.935	.940	.945	.949	.953
.80	.945	.946	.949	.952	.956	.960	.963	.966	.969	.971
.85	.968	.970	.971	.974	.976	.978	.981	.982	.984	.985
.90	.986	.987	.988	.989	.990	.991	.992	.993	.994	.994
.95	.996	.997	.997	.998	.998	.998	.998	.999	.999	.999
1.00	1.000	1.000	1.000	1.000	1.000	1.000	1.000	1.000	1.000	1.000

L \ B	1.2	1.4	1.6	1.8	2.0	2.5	3.0	3.5	4.0	5.0
.05	.524	.575	.617	.652	.681	.737	.777	.806	.828	.861
.10	.545	.594	.634	.667	.695	.748	.786	.814	.836	.867
.15	.573	.619	.656	.687	.714	.764	.799	.825	.846	.875
.20	.605	.648	.682	.711	.735	.782	.814	.838	.857	.884
.25	.641	.679	.711	.737	.759	.801	.831	.853	.870	.894
.30	.678	.712	.741	.764	.784	.822	.848	.868	.883	.905
.35	.715	.746	.771	.792	.809	.843	.866	.884	.897	.917
.40	.753	.780	.801	.820	.835	.864	.884	.899	.911	.928
.45	.790	.812	.831	.847	.860	.884	.902	.915	.924	.939
.50	.825	.844	.860	.872	.883	.904	.918	.929	.937	.949
.55	.858	.873	.886	.897	.906	.922	.934	.943	.949	.959
.60	.888	.901	.911	.919	.926	.939	.948	.955	.960	.968
.65	.915	.925	.933	.939	.944	.954	.961	.966	.970	.976
.70	.939	.946	.952	.956	.960	.967	.972	.976	.979	.983
.75	.959	.964	.968	.971	.974	.978	.982	.984	.986	.989
.80	.975	.978	.981	.983	.984	.987	.989	.991	.992	.993
.85	.987	.989	.990	.991	.992	.993	.994	.995	.996	.997
.90	.995	.996	.996	.997	.997	.997	.998	.998	.998	.999
.95	.999	.999	.999	.999	.999	1.000	1.000	1.000	1.000	1.000

Since the setup is symmetrical, $\bar{\gamma}(\text{raise}_1, A) = \bar{\gamma}(\text{raise}_2, A)$, so that

$$\begin{aligned}\bar{\gamma}(S, A) &= \bar{\gamma}(\text{raise}_1, A) \\ &= 1/(30 \times 15)[7{\cdot}5 \times 15 \times \chi(7{\cdot}5;15) + 22{\cdot}5 \times 15 \times \chi(22{\cdot}5;15)] \\ &= \tfrac{1}{4}\chi(7{\cdot}5;15) + \tfrac{3}{4}\chi(22{\cdot}5;15) \\ &= \tfrac{1}{4} \times 23{\cdot}4 + \tfrac{3}{4} \times 37{\cdot}1 \\ &= 33{\cdot}7\,(\%)^2\end{aligned}$$

This gives an extension variance of:

$$\sigma_e^2 = (2 \times 33{\cdot}7) - 31{\cdot}7 - 34{\cdot}4 = 1{\cdot}3\,(\%)^2$$

This gives a standard error for the estimation of the panel average of 1·14 % Zn. Thus although the original sample standard deviation for 'point' samples was 7 % Zn, two 'samples' within the panel produce a standard error one-sixth of this figure.

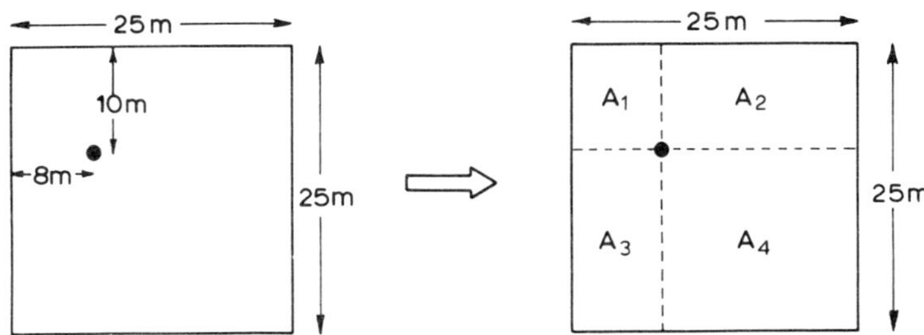

FIG. 4.13. Estimation of panel average from one 'point' sample.

One last brief example: a porphyry copper deposit has been explored by means of vertical boreholes. To simplify the problem, each 'bench' in the deposit is considered separately as a plane, and the borehole intersections as points within the plane. Take a typical small block, 25 m by 25 m, with a borehole passing through it as shown in Fig. 13. Suppose we 'extend' the value of the core intersecting the block to the whole block. Then:

T = average grade of the block
A = 25 metre square panel
T^* = grade of borehole intersection
S = borehole intersection with this bench

Then

$$\sigma_e^2 = 2\bar{\gamma}(S, A) - \bar{\gamma}(S, S) - \bar{\gamma}(A, A)$$

From previous investigation we know that the semi-variogram has a spherical form with a range of influence of 90 m and a sill of $0{\cdot}6\,(\%)^2$ Cu. $\bar{\gamma}(A, A)$ is yet another $F(25, 25)$ function, whose value can be calculated to be $0{\cdot}2145 \times 0{\cdot}6 = 0{\cdot}129\,(\%)^2$ Cu. $\bar{\gamma}(S, S) = 0$ since the sample is supposed to be a point. $\bar{\gamma}(S, A)$ must be calculated with the now familiar (?) manoeuvring, as follows:

$$\bar{\gamma}(S, A) = \bar{\gamma}(\text{point, panel})$$

$= $ (sum of all semi-variogram values between the sample and all points within the panel)/(25×25)

$= $ (sum of all semi-variogram values between the sample and all points in panel A_1
+ the sample and all points in panel A_2
+ the sample and all points in panel A_3
+ the sample and all points in panel A_4)/(25×25)

$$= [8 \times 10 \times H(8,10) + 17 \times 10 \times H(17,10) + 8 \times 15 \times H(8,15) + 17 \times 15 \times H(17,15)]/(25 \times 25)$$

$$= (80 \times 0{\cdot}0703 + 170 \times 0{\cdot}1053 + 120 \times 0{\cdot}0915 + 255 \times 0{\cdot}1248)/(25 \times 25)$$

$$= 0{\cdot}106 \qquad (\%\text{Cu})^2$$

So, finally, the extension variance becomes:

$$\sigma_e^2 = (2 \times 0{\cdot}106) - 0{\cdot}000 - 0{\cdot}129 = 0{\cdot}083\,(\%\text{Cu})^2$$

so that the standard error for the estimation is $0{\cdot}288$ %Cu. The same exercise can be repeated for a sample *outside* the panel, using the same logic as that applied to the one-dimensional problem in Fig. 4.7.

SUMMARY OF MAJOR POINTS

1. When an estimation is performed, an error is made in the prediction.
2. The magnitude of that error is dictated by the structure and type of the deposit, and by the mineral itself. Different minerals within the same deposit may have different structures.

3. The structure can (probably) be described by a semi-variogram model, in the absence of significant trend on the local level.
4. The estimation error variance can be calculated if the semi-variogram model is known. Tables for the spherical model have been supplied. These may also be used as approximations for the linear model.
5. If we use the extension type of estimator, i.e. the arithmetic mean of the samples, then the extension variance may be written:

$$\sigma_e^2 = 2\bar{\gamma}(S, A) - \bar{\gamma}(S, S) - \bar{\gamma}(A, A)$$

That is, the 'reliability' of the estimator depends on three quantities: the relationship of the samples to the area to be estimated; the relationship amongst the samples; and the variation of grades within the area being estimated.

SOME MORE COMPLEX PROBLEMS

Let us return to the original problem posed in Figs. 4.1 and 4.2. We showed that if we used the grade of sample 1 to estimate the *point A*, we obtained a standard error of 25·4 p.p.m. We then introduced the problem of estimating a 60 ft by 30 ft block centred at A. After the examples above, it should be easy enough to calculate that the extension standard error will now be 19·2 p.p.m. [$\bar{\gamma}(S, A) = 356$, $\bar{\gamma}(A, A) = 344$, $\bar{\gamma}(S, S) = 0$], when estimating the average grade of the panel from the single sample point 1. This is over 20% lower than when trying to estimate the central point. The real conclusion is simply that it is easier to estimate the average grade over a block than to specify the grade at a single point. Now, if we consider the arithmetic mean of the 5 point-samples, we obtain the following:

T = panel average grade

A = panel 30 × 60 ft

T^* = average of the 5 samples = 366 p.p.m.

S = 5 point-samples at specified locations.

If we use this estimator to predict the *central point* of the block, the extension standard deviation is 21·8 p.p.m. However, if we estimate the panel, the extension standard deviation reduces to 12·8 p.p.m. Notice that both of these figures are lower than when only considering sample 1. It would appear that, even though the other samples are a lot further away from the *centre* of the block, they are contributing a fair amount of information about the block grade.

To conclude this chapter, an example on a slightly grander scale, on which the reader can exercise his new-found knowledge. For this deposit a simulation has been used, since in that case we *know* the semi-variogram and the value at every point within the deposit. This enables us to compare estimates with 'actual' values—a situation which is rare to the point of extinction in the real world. It also enables us to produce a set of samples on

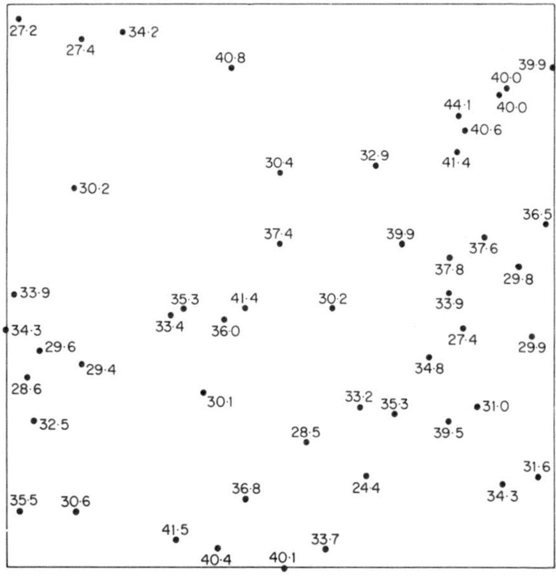

FIG. 4.14. A set of random samples taken from a simulated iron ore deposit.

any given sampling scheme proposed. The simulated deposit is a low grade sedimentary iron ore, with an overall average of about 35 % Fe, a standard deviation of 5 % Fe, a range of influence of 100 m and a sill of 25 $(\%)^2$ Fe—obviously. The semi-variogram is spherical (yet again) with no nugget effect. An area 400 metres square has been simulated and a set of 50 samples taken from it at random. The positions and values of these are shown in Fig. 4.14 and Table 4.5. The initial estimation, at the pre-feasibility stage, is to be done on 50 m by 50 m blocks. For this first example, each block has been allocated the average grade of all interior samples. Where a sample falls on the edge it has been allocated to both blocks. Figure 4.15 shows the

TABLE 4.5. RANDOM SAMPLES TAKEN FROM A SIMULATED IRON ORE EXAMPLE

Easting	*Northing*	*% Fe*	*Easting*	*Northing*	*% Fe*
0	170	34·3	5	195	33·9
10	40	35·5	20	105	32·5
15	135	28·6	25	155	29·6
55	145	29·4	50	40	30·6
125	20	41·5	155	15	40·4
175	50	36·8	145	125	30·1
120	180	33·4	130	185	35·3
160	175	36·0	175	185	41·4
240	185	30·2	220	90	28·5
260	115	33·2	205	0	40·1
235	15	33·7	265	65	24·4
365	60	34·3	390	65	31·6
285	110	35·3	325	105	39·5
345	115	31·0	310	150	34·8
335	170	27·4	385	165	29·9
325	195	33·9	325	220	37·8
350	235	37·6	375	215	29·8
290	230	39·9	200	230	37·4
10	390	27·2	55	375	27·4
85	380	34·2	395	245	36·5
50	270	30·2	165	355	40·8
200	280	30·4	270	285	32·9
400	355	39·9	365	340	40·0
360	335	40·0	330	320	44·1
335	310	40·6	330	290	41·4

estimated value for each block. Blocks without internal sampling have been shaded in. The upper figure shown in each block is the estimator T^*, and the lower is the extension standard deviation. Since *this* deposit is actually Normally distributed, 95 % confidence limits would be given (approximately) by $T^* \pm 2\sigma_e$. For comparison, the 'true' average grade for each block is shown in Fig. 4.16. It can be seen that in most of the 37 estimated blocks, the true value is within the 95 % confidence interval. Four or five blocks lie just outside the interval, and three or four are considerably outside. This is a little higher than would be expected, since we would only expect about two blocks to lie outside a 95 % interval. However, if we consider the 99 % confidence interval (3 standard deviations) only one block is significantly outside the interval, i.e. the lower left-hand block of the area (south-west corner).

27·2 3·2	30·8 1·5		40·8 3·3				39·9 4·0
						42·4 2·2	40·0 2·7
30·2 3·5	30·2 3·5		30·4 3·5	30·4 3·5	32·9 2·5	41·4 3·1	
			37·5 3·5	37·5 3·5	39·9 2·8	37·7 1·9	34·6 1·3
32·6 1·9		34·4 1·9	38·7 1·8	30·2 2·9		32·1 1·2	29·9 2·6
30·5 1·7	29·4 3·7	30·1 3·0			34·2 1·9	35·1 1·4	
			36·8 3·5	28·5 2·8	24·4 2·7		32·9 2·0
33·1 2·2	30·6 3·7	41·5 2·2	38·6 2·0	36·9 2·3			

FIG. 4.15. Estimates of block values formed by averaging all interior samples in the iron ore deposit, and the corresponding extension standard deviations.

36·2	35·0	43·0	44·2	37·5	38·5	40·5	38·7
36·1	28·2	38·4	36·3	30·8	35·2	41·6	38·5
39·5	25·1	34·2	33·5	29·3	36·0	39·2	32·8
39·5	31·2	38·5	40·0	36·5	38·2	37·4	32·8
33·1	32·4	34·5	38·5	36·2	33·9	30·0	30·9
35·9	36·9	36·5	33·6	34·9	34·5	34·2	33·6
35·0	34·1	37·5	32·7	27·7	31·4	35·5	34·2
41·3	31·6	38·8	37·9	33·6	29·8	35·8	35·0

FIG. 4.16. 'Actual' average values within each block in the simulated iron ore deposit.

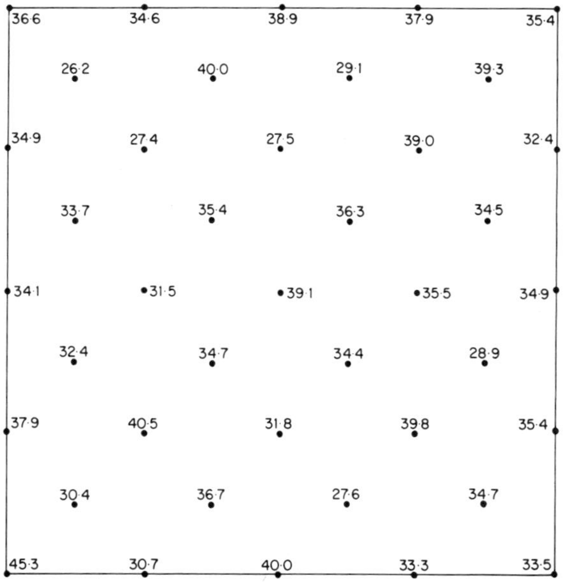

FIG. 4.17. A set of samples taken on a regular grid from the simulated iron ore deposit.

31·4	30·4	37·3	39·5	34·0	33·5	38·6	37·4
30·6	36·8	33·7	33·7	28·3	34·0	39·1	35·8
34·3	30·6	31·4	31·4	31·9	37·7	36·8	33·4
33·9	32·6	33·4	37·2	37·7	35·9	35·0	34·7
33·2	31·9	33·1	36·9	36·8	35·0	32·2	31·9
35·1	36·4	37·6	33·3	33·1	37·1	34·3	32·2
34·1	35·4	38·6	34·2	29·7	33·7	37·2	35·1
37·8	30·5	33·7	38·3	33·8	30·5	34·0	34·1

FIG. 4.18. Estimated block values from regular samples.

For comparison, Fig. 4.17 shows a set of samples taken on a regular grid from the same 'deposit'. In this case, each 50 m block has two samples in opposing corners. Using the average of these two samples to estimate the block results in an extension standard deviation of 2·5 % Fe, Fig. 4.18 shows the estimated values in each block. It can be seen that although five or six blocks lie outside the 95 % confidence interval, not one lies more than 2·25 standard deviations from the true value.

Having exhausted the possibilities of extension in idealised circumstances, let us move on to some more interesting situations.

CHAPTER 5

Kriging

Let us turn, now, to a much more common, and probably more realistic, approach to the estimation of 'local' values. Until now we have considered only the operation of averaging all the local samples and applying this value as the estimate of the area under consideration. There are cases in which it would not seem sensible to weight all the samples equally, since some will be a great distance from the 'unknown' area A, whilst others will be much closer to it—if not inside. It would seem more sensible to use a weighted average of the sample values, with the 'closer' samples having more weight. The new estimator will be of the form:

$$T^* = w_1 g_1 + w_2 g_2 + w_3 g_3 + \cdots + w_n g_n$$

where the weightings sum to 1. If this condition is met, *and* there is no trend (locally) then T^* is an unbiased estimator. This means that over a lot of estimations the average error will be zero. This type of estimator is called a 'linear' estimator because it is a linear combination of the sample values. The arithmetic mean is simply a special case where all of the weights are equal. It can be shown that the estimation variance for the general 'unbiased linear' estimator is:

$$\sigma_\varepsilon^2 = 2\sum_{i=1}^{n} w_i \bar{\gamma}(S_i, A) - \sum_{i=1}^{n}\sum_{j=1}^{n} w_i w_j \bar{\gamma}(S_i, S_j) - \bar{\gamma}(A, A)$$

Where we previously calculated $\bar{\gamma}(S, A)$ which was the average semi-variogram between each sample and the unknown area, we now form a weighted average for each individual sample with the area A, $\bar{\gamma}(S_i, A)$, in the same way as the actual estimator. For example, if we have n samples taken around the area A, the first term in the variance becomes:

$$\sum_{i=1}^{n} w_i\bar{\gamma}(S_i, A) = w_1\bar{\gamma}(\text{point}_1, A) + w_2\bar{\gamma}(\text{point}_2, A) + w_3\bar{\gamma}(\text{point}_3, A)\ldots\text{etc.}$$

The last term in the variance, $\bar{\gamma}(A, A)$, does not change its form, since we have only changed the form of the estimator, not the area being estimated. The term $\bar{\gamma}(S, S)$ which previously measured the variation in values between the samples, must now take into account the different weights associated with each sample. Thus, if we take, e.g., sample 4 we must remember that it has a weight w_4. If we take sample 4 with, say, sample 2, then we must include both weights w_2 and w_4, so that the term becomes

$$w_2 w_4 \bar{\gamma}(S_2, S_4)$$

To form the equivalent $\bar{\gamma}(S, S)$, then, each $\bar{\gamma}(S_i, S_j)$ is multiplied by the corresponding $w_i w_j$ before being added into the sum.

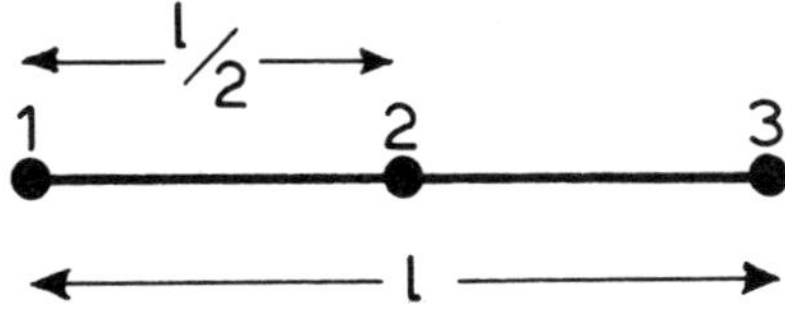

FIG. 5.1. Three samples to be used to estimate the line segment.

As a simple example, let us consider the setup in Fig. 5.1. We calculated the extension variance for this setup in Chapter 4. The arithmetic mean gave an extension variance of $pl/18$ when we used a linear semi-variogram with the form $\gamma(h) = ph$. Now suppose we allocate a set of weights to these three samples instead of treating them all the same. Let us, for this example, put three-quarters of the 'weight' at the central point, and an eighth on each end point. The 'new' estimator is therefore:

$$T^* = w_1 g_1 + w_2 g_2 + w_3 g_3 = \tfrac{1}{8}g_1 + \tfrac{3}{4}g_2 + \tfrac{1}{8}g_3$$

$$A = \text{the length } l$$

$$S = 3 \text{ point-samples.}$$

The reliability of this estimator will be given by the general form of σ_ε^2. The term $\bar{\gamma}(A, A)$ is given by the function $F(l)$, by definition, which for the linear semi-variogram takes the form $pl/3$. The central term—the 'within samples' term—is:

$$\sum\sum w_i w_j \bar{\gamma}(S_i, S_j) = w_1[w_1\bar{\gamma}(S_1, S_1) + w_2\bar{\gamma}(S_1, S_2) + w_3\bar{\gamma}(S_1, S_3)] + w_2[w_1\bar{\gamma}(S_2, S_1) + w_2\bar{\gamma}(S_2, S_2) + w_3\bar{\gamma}(S_2, S_3)] + w_3[w_1\bar{\gamma}(S_3, S_1) + w_2\bar{\gamma}(S_3, S_2) + w_3\bar{\gamma}(S_3, S_3)]$$

where each sample is taken with each sample in turn—including itself—and the combination is multiplied by both weights. Since the samples in this case are all points, all the $\bar{\gamma}(S_i, S_j)$ terms can be found directly from the semi-variogram $\gamma(h)$. This gives:

$$\sum\sum w_i w_j \bar{\gamma}(S_i, S_j) = w_1[w_1\gamma(0) + w_2\gamma(l/2) + w_3\gamma(l)] + w_2[w_1\gamma(l/2) + w_2\gamma(0) + w_3\gamma(l/2)] + w_3[w_1\gamma(l) + w_2\gamma(l/2) + w_3\gamma(0)]$$

Since the semi-variogram model is linear, this becomes:

$$\sum\sum w_i w_j \bar{\gamma}(S_i, S_j) = \frac{1}{8}\left(\frac{3}{4}p\frac{l}{2} + \frac{1}{8}pl\right) + \frac{3}{4}\left(\frac{1}{8}p\frac{l}{2} + \frac{1}{8}p\frac{l}{2}\right) + \frac{1}{8}\left(\frac{1}{8}pl + \frac{3}{4}p\frac{l}{2}\right)$$

$$= pl\frac{1}{8}\left(\frac{3}{8} + \frac{1}{8} + \frac{3}{8} + \frac{3}{8} + \frac{1}{8} + \frac{3}{8}\right)$$

$$= \frac{7}{32}pl$$

This leaves only the first term—the between sample and area term—to be evaluated. This is:

$$\sum w_i \bar{\gamma}(S_i, A) = w_1\bar{\gamma}(S_1, A) + w_2\bar{\gamma}(S_2, A) + w_3\bar{\gamma}(S_3, A)$$

Sample 1 is at one end of the length l, so that $\bar{\gamma}(S_1, A) = \chi(l)$. Similarly $\bar{\gamma}(S_3, A) = \chi(l)$. Sample 2 is the central point of the length l, so that $\bar{\gamma}(S_2, A) = \chi(l/2)$. For the linear model $\chi(l) = pl/2$, so that

$$\sum w_i \bar{\gamma}(S_i, A) = w_1\chi(l) + w_2\chi(l/2) + w_3\chi(l)$$

$$= \frac{1}{8}p\frac{l}{2} + \frac{3}{4}p\frac{l}{4} + \frac{1}{8}p\frac{l}{2}$$

$$= \frac{5}{16}pl$$

Putting all these together, gives an estimation variance of:

$$\sigma_\varepsilon^2 = 2\frac{5}{16}pl - \frac{7}{32}pl - \frac{1}{3}pl$$

$$= \frac{pl}{96}(60 - 21 - 32) = \frac{7}{96}pl = 0{\cdot}0729\,pl$$

The specified set of weights, $(\frac{1}{8}, \frac{3}{4}, \frac{1}{8})$ when used with a linear semi-variogram model give an estimation variance of $0{\cdot}0729\,pl$. The arithmetic mean of the samples gave an extension variance of $pl/18 = 0{\cdot}0556\,pl$, which is three-quarters of the size of the estimation variance above. It is fairly obvious in this case that the specified weighted average gives a worse estimator than simply using the arithmetic mean.

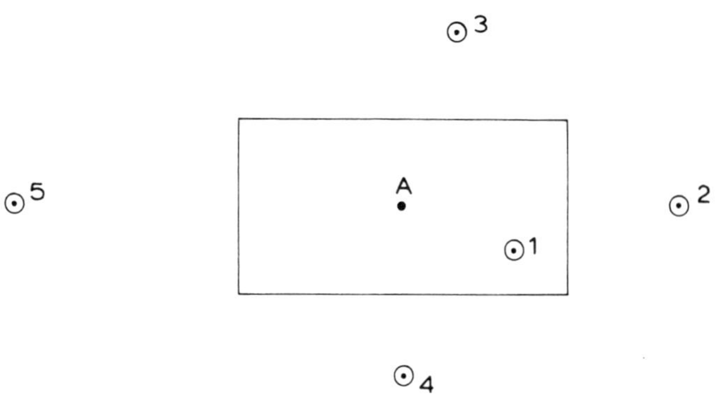

FIG. 5.2. Estimation of the block value is required from the five scattered samples—uranium example.

As a two-dimensional example, let us return to the ubiquitous uranium example as illustrated in Fig. 5.2. We shall allocate weights to each sample according to its distance from the centre of the block A. Table 5.1 shows the 'inverse distance' calculation for the weights in this example. The estimator T^* becomes:

$$T^* = w_1g_1 + w_2g_2 + w_3g_3 + w_4g_4 + w_5g_5$$

$$= 0{\cdot}319 \times 400 + 0{\cdot}137 \times 380 + 0{\cdot}217 \times 450 + 0{\cdot}229 \times 280 + 0{\cdot}098 \times 320$$

$$= 372{\cdot}8 \text{ p.p.m. } U_3O_8$$

TABLE 5.1. CALCULATION OF INVERSE DISTANCE WEIGHTINGS FOR HYPOTHETICAL URANIUM ESTIMATION PROBLEM

Sample number	*Distance from centre (ft)*	*Inverse distance*	*Corrected weight*
1	21·54	0·0464	0·319
2	50·00	0·0200	0·137
3	31·62	0·0316	0·217
4	30·00	0·0333	0·229
5	70·00	0·0143	0·098
Total		0·1457	1·000

The three terms which go into the variance are:

$$\sum w_i \bar{\gamma}(S_i, A) = 0{\cdot}319\,\bar{\gamma}(S_1, A) + 0{\cdot}137\,\bar{\gamma}(S_2, A) + 0{\cdot}217\,\bar{\gamma}(S_3, A) + 0{\cdot}229\,\bar{\gamma}(S_4, A) + 0{\cdot}098\,\bar{\gamma}(S_5, A)$$

The individual values of $\bar{\gamma}(S_1, A)$ and so on are the same as those evaluated when considering the extension variance, and are equal to 356·7, 572·4, 456·9, 446·8 and 696·1 respectively. Thus the first term in the variance is equal to 461·9 (p.p.m.)^2—as opposed to 505·8 (p.p.m.)^2 in the extension variance. The second term in the variance of the weighted mean turns out to be 441·2 (p.p.m.)^2. In the extension case it was 504·7 (p.p.m.)^2. The final term in both variances is 344·0 (p.p.m.)^2 as given by the auxiliary function $F(l, b)$. Putting these figures together gives us an estimation variance for the inverse distance estimator of 138·6 (p.p.m.)^2. This is equivalent to a 'standard error' of 11·8 p.p.m. Thus, using the inverse distance weights we obtain an estimate of 372·8 p.p.m. U_3O_8 for the panel, and we can say that this has a standard error of 11·8 p.p.m., which has been derived from our knowledge of the semi-variogram of this deposit. If we are willing to make the additional assumption of Normality of the errors, we could say that a 95% confidence interval for the 'true' value of the panel is (349 p.p.m., 396 p.p.m.). This should be compared to the extension estimate of 366 p.p.m., with a standard error of 12·8 p.p.m., and a 95% confidence interval of (340 p.p.m., 392 p.p.m.). It can quickly be seen that the inverse distance estimation produces a (slightly) more accurate result than the arithmetic mean—this seems quite sensible.

Suppose we change the situation slightly. Suppose that sample 3 was not north of the block but south, i.e. northing 2310, as in Fig. 5.3. The inverse distance weights are unchanged, because they depend on the distance

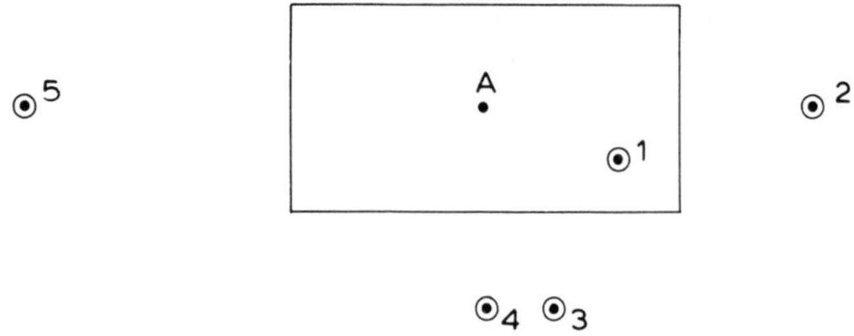

FIG. 5.3. Sample 3 is now located South of the block to be estimated.

between the sample and the block centre. However, the estimation standard error rises sharply to 14·3 p.p.m., indicating a loss in accuracy. The change in the estimation variance is caused solely by a decrease in the size of the 'within samples' term—the amount of information contained in the sample set—since the other two terms remain the same as before. In actual figures the term falls from 441·2 (p.p.m.)^2 to 376·2 (p.p.m.)^2. It does not really seem very sensible to use the same weights for Fig. 5.3 as we did for Fig. 5.2, since sample 3 now gives us a lot less 'information' than it did previously. We could suggest a new set of weights and calculate the estimation variance. If this were smaller than the 'inverse distance' set, we could say our 'new' estimator was (in some sense) better than the inverse distance one. Alternatively, we could use a different method to produce the weights—inverse distance squared, for example. It would seem a lot more desirable to find a direct method of producing the 'best' estimate, given our knowledge of the deposit.

We have decided to use a 'linear' type of estimator—a weighted average of the sample values. We know that it is an unbiased estimator if the weights add up to 1. There is an infinite number of such linear unbiased estimators, so we will search for the 'best' one, and we will define 'best' as 'having the smallest estimation variance'. The expression for the estimation variance depends on three things: the basic geometry of samples and unknown area, the form of the semi-variogram, and the weighting allocated to each sample. For any given setup the variance can only be changed by altering the values of the *weights*. Thus we wish to minimise the estimation variance with respect to the weights. The variance is a simple (?) function of the weights, so to minimise it we must differentiate and set the differential equal to zero:

i.e. $$\frac{\partial \sigma_\varepsilon^2}{\partial w_i} = 0 \qquad i = 1, 2, 3, 4 \ldots n$$

This will provide n equations in n unknowns $(w_1, w_2, w_3, w_4 \ldots w_n)$. These weights will provide an estimator which has the minimum value of the estimation variance. However, they will not necessarily add up to *one*. There is nothing in the above system of equations that constrains the weights in this way. Effectively, we also need to satisfy the equation $\Sigma w_i = 1$. Thus, to obtain the 'Best Linear Unbiased Estimator' we actually have to satisfy $(n + 1)$ equations. However, we only have n unknowns, so far, which is not a very desirable condition. To rectify this we must introduce another unknown, in the form of a Lagrangian Multiplier, to balance up the system. Therefore, instead of minimising the estimation variance, we actually minimise:

$$\sigma_\varepsilon^2 - \lambda(\Sigma w_i - 1)$$

with respect to $w_1, w_2, w_3, w_4 \ldots w_n$, and λ. This last produces the equation:

$$\Sigma w_i - 1 = 0$$

as is required.

After the differentiation has been performed and the equations tidied up, the following system results:

$$\begin{array}{l}
w_1\bar{\gamma}(S_1, S_1) + w_2\bar{\gamma}(S_1, S_2) + w_3\bar{\gamma}(S_1, S_3) + \cdots + w_n\bar{\gamma}(S_1, S_n) + \lambda = \bar{\gamma}(S_1, A) \\
w_1\bar{\gamma}(S_2, S_1) + w_2\bar{\gamma}(S_2, S_2) + w_3\bar{\gamma}(S_2, S_3) + \cdots + w_n\bar{\gamma}(S_2, S_n) + \lambda = \bar{\gamma}(S_2, A) \\
w_1\bar{\gamma}(S_3, S_1) + w_2\bar{\gamma}(S_3, S_2) + \cdots \quad \cdots \quad \cdots + w_n\bar{\gamma}(S_3, S_n) + \lambda = \bar{\gamma}(S_3, A) \\
\quad \cdots \qquad \cdots \qquad\qquad\qquad \cdots \qquad \cdots \\
\quad \cdots \qquad \cdots \qquad\qquad\qquad \cdots \qquad \cdots \\
w_1\bar{\gamma}(S_n, S_1) + w_2\bar{\gamma}(S_n, S_2) + w_3\bar{\gamma}(S_n, S_3) + \quad + w_n\bar{\gamma}(S_n, S_n) + \lambda = \bar{\gamma}(S_n, A) \\
\quad w_1 \quad + \quad w_2 \quad + \quad w_3 \quad + \quad + \quad w_n \qquad = \quad 1
\end{array}$$

Although this looks slightly fearsome in its complexity, if you look a little closer, you will find that most of the elements are (I hope) by now familiar to you. Consider the first equation. The right hand side merely requires the average semi-variogram value between sample 1 and the unknown area. The left hand side contains the $n + 1$ unknowns, w_i and λ, and the average semi-variogram value between sample 1 and each of the other samples in turn. All of these $\bar{\gamma}$ terms are identical to those which we would work out for an extension or an estimation variance.

The second equation is identical to the first except that it is sample 2 which is present right along the equation. The third has sample 3 all the way along, and so on until the nth equation with S_n all the way along. Finally, we have the necessary condition for the sum of the weights. The solution to this set of equations will produce a set of weights giving the 'Best Linear Unbiased Estimator'—sometimes referred to as BLUE. This process was named Kriging by Georges Matheron after Danie Krige, who has done a tremendous amount of empirical work on weighted averages. Pronunciation of the word is a controversial topic—I personally prefer 'kridging' (as in bridging). The system of equations is generally referred to as the kriging system, and the estimator produced is the kriging estimator. The variance of the kriging estimator could be found by substitution of the weights into the general estimation variance equation. However, it can be shown that the kriging variance can be written as:

$$\sigma_k^2 = \sum w_i \bar{\gamma}(S_i, A) + \lambda - \bar{\gamma}(A, A)$$

KRIGING EXAMPLES

Earlier in this chapter we took the setup in Fig. 5.1, a semi-variogram of the form $\gamma(h) = ph$, and allocated a set of weights to each of the three samples. We showed that in that case the estimation variance was $0{\cdot}0729\,pl$. Now let us see if we can find the 'best' set of weights using the kriging system. Since we have three weights, there are four equations which in the general form are:

$$\begin{aligned}
w_1\bar{\gamma}(S_1,S_1) + w_2\bar{\gamma}(S_1,S_2) + w_3\bar{\gamma}(S_1,S_3) + \lambda &= \bar{\gamma}(S_1,A)\\
w_1\bar{\gamma}(S_2,S_1) + w_2\bar{\gamma}(S_2,S_2) + w_3\bar{\gamma}(S_2,S_3) + \lambda &= \bar{\gamma}(S_2,A)\\
w_1\bar{\gamma}(S_3,S_1) + w_2\bar{\gamma}(S_3,S_2) + w_3\bar{\gamma}(S_3,S_3) + \lambda &= \bar{\gamma}(S_3,A)\\
w_1 \qquad\qquad + w_2 \qquad\qquad + w_3 \qquad\qquad &= 1
\end{aligned}$$

and the kriging variance would be:

$$\sigma_k^2 = w_1\bar{\gamma}(S_1,A) + w_2\bar{\gamma}(S_2,A) + w_3\bar{\gamma}(S_3,A) + \lambda - \bar{\gamma}(A,A)$$

The right hand side of the equations are $\bar{\gamma}(S_1, A)$, $\bar{\gamma}(S_2, A)$ and $\bar{\gamma}(S_3, A)$. $\bar{\gamma}(S_1, A)$ is the average semi-variogram value between a line of length l and a point on the end of it. This is the definition of $\chi(l)$. So is $\bar{\gamma}(S_3, A)$. $\bar{\gamma}(S_2, A)$ is the average semi-variogram between the line and a central point. This example was tackled in Chapter 4 and found to be $\chi(l/2)$. Thus we have the right hand side of the equations and most of the variance. The left hand side of the equations is made up of the individual sample-with-sample terms. Since all of the samples in this case are supposed to be points, all of these 'left hand side' relationships are given by the semi-variogram model. It remains only to calculate the distances between the pairs and use the model to produce the terms. The diagonal terms, $\bar{\gamma}(S_1, S_1)$, $\bar{\gamma}(S_2, S_2)$ and $\bar{\gamma}(S_3, S_3)$ are all zero, since $\gamma(0) = 0$ by definition, $\bar{\gamma}(S_1, S_2)$ is equal to $\bar{\gamma}(S_2, S_1)$, and is $\gamma(l/2)$. So is $\bar{\gamma}(S_2, S_3)$ and $\bar{\gamma}(S_3, S_2)$. $\bar{\gamma}(S_1, S_3)$ is $\gamma(l)$ as is $\bar{\gamma}(S_3, S_1)$. Finally, $\bar{\gamma}(A, A)$ is $F(l)$ by definition. Putting these together gives:

$$\begin{aligned} w_2\gamma(l/2) + w_3\gamma(l) + \lambda &= \chi(l) \\ w_1\gamma(l/2) + w_3\gamma(l/2) + \lambda &= \chi(l/2) \\ w_1\gamma(l) + w_2\gamma(l/2) + \lambda &= \chi(l) \\ w_1 + w_2 + w_3 &= 1 \end{aligned}$$

and the kriging variance is

$$\sigma_k^2 = w_1\chi(l) + w_2\chi(l/2) + w_3\chi(l) + \lambda - F(l)$$

Up until this point the system does not depend on the actual model for the semi-variogram. Our model was $\gamma(h) = ph$ and for this example we will take $p = 4$. For the linear semi-variogram, $\chi(l) = pl/2$, so in this case $\chi(l) = 2l$. Similarly, $F(l) = 4l/3$. Substituting in the above system:

$$\begin{aligned} 2lw_2 + 4lw_3 + \lambda &= 2l \quad (1) \\ 2lw_1 + 2lw_3 + \lambda &= l \quad (2) \\ 4lw_1 + 2lw_2 + \lambda &= 2l \quad (3) \\ w_1 + w_2 + w_3 + &= 1 \quad (4) \end{aligned}$$

and

$$\sigma_k^2 = 2lw_1 + lw_2 + 2lw_3 + \lambda - \tfrac{4}{3}l$$

Adding equations (1) and (3) gives

$$4lw_1 + 4lw_2 + 4lw_3 + \lambda = 4l$$

whilst equation (4) gives

$$w_1 + w_2 + w_3 = 1$$

These two together show that *in this case* $\lambda = 0$. Thus we can eliminate λ from the first three equations, and divide all of them through by l. This suggests that the results—the final values of the weights—do not rely on the length being estimated. We have then:

$$2w_2 + 4w_3 = 2 \quad (5)$$

$$2w_1 + 2w_3 = 1 \quad (6)$$

$$4w_1 + 2w_2 = 2 \quad (7)$$

Subtracting equation (5) from equation (7) gives:

$$4w_1 - 4w_3 = 0 \qquad \text{i.e. } w_1 = w_3$$

so that equation (6) gives:

$$4w_1 = 1 \qquad \text{i.e. } w_1 = \tfrac{1}{4}$$

Therefore, $w_3 = \frac{1}{4}$ and $w_2 = \frac{1}{2}$. The 'optimal' set of weights for the problem in Fig. 5.1 is therefore $(\frac{1}{4}, \frac{1}{2}, \frac{1}{4})$ and this estimator gives a kriging variance of:

$$\sigma_k^2 = 2\frac{1}{4}l + \frac{1}{2}l + 2\frac{1}{4}l + 0 - \frac{4}{3}l = \frac{l}{6}$$

The final result is therefore that:

the BLUE has weights of $(\frac{1}{4}, \frac{1}{2}, \frac{1}{4})$ and a kriging variance of $l/6$.

In our previous studies of this particular setup, we found that the arithmetic mean gave an extension variance of $pl/18$ and that the set of weights $(\frac{1}{8}, \frac{3}{4}, \frac{1}{8})$ gave $7pl/96$. To match the example above we should set $p = 4$, so that the variances become $2l/9$ and $7l/24$ respectively. The kriging procedure has improved over the arithmetic mean by about 25 %, this being the difference in magnitude between the extension variance and the kriging variance. The spurious set of weights which we tried earlier in this chapter give a variance almost twice that of the optimal estimator. Note that in this case, since we have used a *linear* semi-variogram model, the set of weights is independent of the length being estimated, but the variance is directly proportional to it. For an exercise, see if you can show that the set of weights is also independent of the *slope* of the semi-variogram, p. Show also that the kriging variance is directly proportional to p, which seems only sensible.

TWO-DIMENSIONAL EXAMPLE

Now let us return to the familiar uranium example shown in Fig. 5.2. The data—position co-ordinates and grade values—were given in Table 4.1. We have five samples, so there will be six equations in the kriging system. The $\bar{\gamma}(S_i, S_j)$ terms on the left hand side of the equations are all 'point-with-point' relationships, and can be evaluated directly from the semi-variogram model. The model was spherical, with a range of influence of 100 ft, a sill of 700 (p.p.m.)^2 and a nugget effect of 100 (p.p.m.)^2. The $\bar{\gamma}(S_i, A)$ terms are all 'point-with-panel' relationships and hence are simple combinations of the $H(l, b)$ auxiliary function. The $\bar{\gamma}(A, A)$ term is, of course, $F(l, b)$.

The kriging system turns out to be:

$$
\begin{aligned}
&\phantom{415{\cdot}5w_1 +{}} 415{\cdot}5w_2 + 491{\cdot}4w_3 + 403{\cdot}0w_4 + 790{\cdot}5w_5 + \lambda = 356{\cdot}7 \\
&415{\cdot}5w_1 + 581{\cdot}3w_3 + 642{\cdot}9w_4 + 800{\cdot}0w_5 + \lambda = 572{\cdot}4 \\
&491{\cdot}4w_1 + 581{\cdot}3w_2 + 659{\cdot}9w_4 + 778{\cdot}8w_5 + \lambda = 456{\cdot}9 \\
&403{\cdot}0w_1 + 642{\cdot}9w_2 + 659{\cdot}9w_3 + 745{\cdot}1w_5 + \lambda = 446{\cdot}8 \\
&790{\cdot}5w_1 + 800{\cdot}0w_2 + 778{\cdot}8w_3 + 745{\cdot}1w_4 + \lambda = 696{\cdot}1 \\
&w_1 + w_2 + w_3 + w_4 + w_5 = 1
\end{aligned}
$$

and the kriging variance would be:

$$\sigma_k^2 = 356{\cdot}7w_1 + 572{\cdot}4w_2 + 456{\cdot}9w_3 + 446{\cdot}8w_4 + 696{\cdot}1w_5 + \lambda - 344{\cdot}0$$

Solving this system of equations (by computer) gives:

$$
\begin{aligned}
w_1 &= 0{\cdot}346 \\
w_2 &= 0{\cdot}023 \\
w_3 &= 0{\cdot}269 \\
w_4 &= 0{\cdot}234 \\
w_5 &= 0{\cdot}127 \\
\lambda &= 19{\cdot}72 \\
T^* &= 376{\cdot}5\ \text{p.p.m.} \\
\sigma_k &= 11{\cdot}3\ \text{p.p.m.}
\end{aligned}
$$

This kriging standard deviation compares favourably with the standard error of the 'inverse distance' weights. The main difference in the weighting is perhaps a surprising one at first sight. Sample 2 is allocated an almost zero weight. In fact, it receives only 20% of the weight on sample 5, which is considerably farther from the block centre. This is because the kriging system automatically takes account of the relationship amongst the samples. Sample 2 provides little extra information about the block in the presence of sample 1, and to a lesser extent samples 3 and 4. To illustrate this point further, let us consider the situation in Fig. 5.3, where sample 3 was moved to the south of the block. The kriging system now becomes:

$$
\begin{aligned}
415{\cdot}5w_2 + 348{\cdot}8w_3 + 403{\cdot}0w_4 + 790{\cdot}5w_5 + \lambda &= 356{\cdot}7\\
415{\cdot}5w_1 + 581{\cdot}3w_3 + 642{\cdot}9w_4 + 800{\cdot}0w_5 + \lambda &= 572{\cdot}4\\
348{\cdot}8w_1 + 581{\cdot}3w_2 + 204{\cdot}6w_4 + 778{\cdot}8w_5 + \lambda &= 456{\cdot}9\\
403{\cdot}0w_1 + 642{\cdot}9w_2 + 204{\cdot}6w_3 + 745{\cdot}1w_5 + \lambda &= 446{\cdot}8\\
790{\cdot}5w_1 + 800{\cdot}0w_2 + 778{\cdot}8w_3 + 745{\cdot}1w_4 + \lambda &= 696{\cdot}1\\
w_1 + w_2 + w_3 + w_4 + w_5 &= 1
\end{aligned}
$$

Note that the only difference between this and the previous system is in row 3 and column 3 on the left hand side. The right hand side is unchanged since we have not changed the relationship of each individual sample to the panel. The new weights are:

$$
\begin{aligned}
w_1 &= 0{\cdot}440\\
w_2 &= 0{\cdot}089\\
w_3 &= 0{\cdot}062\\
w_4 &= 0{\cdot}224\\
w_5 &= 0{\cdot}185 \qquad \lambda = 61{\cdot}58\\
T^* &= 359{\cdot}6\,\text{p.p.m.} \qquad \sigma_k = 13{\cdot}5\,\text{p.p.m.}
\end{aligned}
$$

The weight on the third sample is now less than that of sample 2. The main movement in weight has been towards the 'northern' samples, 1, 2 and 5,

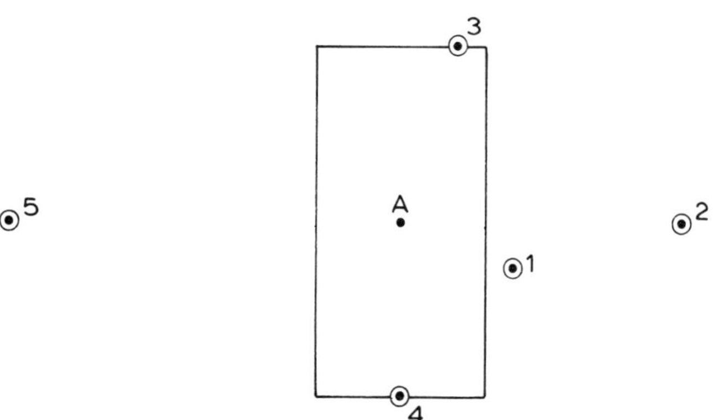

FIG. 5.4. Block to be estimated is rotated through 90°.

although the largest increase is, naturally, in sample 1. The really striking change is in the value of the *estimator* which has dropped by 17 p.p.m.

As a third example let us consider the same sampling situation as before, but now with the block rotated through 90 degrees, as in Fig. 5.4. The kriging system for this setup has the same left hand side as the first situation in Fig. 5.2. However, all of the terms on the right hand side have changed, to:

377·6 599·7 430·2 414·0 720·3

The weights allocated to the samples change drastically:

$$
\begin{aligned}
w_1 &= 0{\cdot}275\\
w_2 &= 0{\cdot}006\\
w_3 &= 0{\cdot}324\\
w_4 &= 0{\cdot}306\\
w_5 &= 0{\cdot}089\\
\lambda &= 20{\cdot}29
\end{aligned}
$$

The estimator T^* takes the value 371·9 p.p.m., and has a standard error of 10·7 p.p.m. Sample 2 has been completely screened out by sample 1, and the influence of sample 5 has also declined somewhat. The biggest surprise, perhaps, is that sample 1 no longer has anything like the importance it had in the previous two examples. There is no method other than kriging in which this shift in 'information value' can be quantified and utilised.

SUMMARY OF MAJOR POINTS

(1) We can evaluate the accuracy of *any* linear estimator if we have a model for the semi-variogram.

(2) We can produce the minimum variance unbiased linear estimator using the kriging technique, if we have a model for the semi-variogram.

The standard litany of the advantages of kriging can be found in numerous publications. The points of major importance are:

(a) Given the basic assumptions, no trend, and a model for the semi-variogram, kriging *always* produces the Best Linear Unbiased Estimator.

(b) If the proper models are used for the semi-variogram, and the system is set up correctly, there is *always* a unique solution to the kriging system.

(c) If you try to 'estimate' the value at a location which has been sampled, the kriging system will return the sample value as the estimator, and a kriging variance of zero. In other words, you already *know* that value. This is usually referred to as an 'exact interpolator'.

(d) If you have regular sampling, and hence the same sampling/block setup at many different positions within the deposit, it is not necessary to recalculate the kriging system each time.

SIMULATED IRON ORE EXAMPLE

To round out this chapter, let us return to the simulated iron ore deposit mentioned at the end of Chapter 4. Two sets of samples had been taken. The 50 'random' samples were shown in Fig. 4.14 and the regular grid in Fig. 4.17.

27·5 1·9	31·7 1·4	36·9 3·0	38·0 2·8	37·9 4·4	37·9 4·5	40·1 4·1	40·5 3·0
29·0 4·0	31·7 3·7	35·9 3·8	36·5 2·8	35·2 3·6	37·3 3·3	41·4 1·7	39·8 1·9
30·5 2·9	31·7 2·9	33·8 4·1	33·6 2·7	32·1 2·2	35·2 1·7	40·0 1·9	38·3 2·7
32·4 3·0	32·0 3·6	34·1 3·5	37·9 2·4	34·9 2·2	36·8 1·9	37·6 1·1	33·5 1·1
31·7 1·6	31·3 2·8	33·6 1·3	37·1 1·4	32·7 2·0	33·0 2·5	31·6 1·0	29·4 1·5
30·3 1·1	31·1 2·7	31·0 2·3	31·8 2·7	31·2 2·6	34·4 1·4	34·5 1·2	30·8 2·5
33·3 2·2	32·7 3·4	34·9 3·4	33·6 2·3	29·0 1·7	29·1 1·6	35·2 2·5	32·5 1·5
33·4 2·0	33·7 2·6	39·4 1·6	38·8 1·5	33·8 1·6	30·2 3·1	32·9 4·0	32·2 3·5

FIG. 5.5. Simulated iron ore deposit—block averages kriged from the set of random samples and the corresponding kriging standard deviations—50 metre blocks.

The regular grid actually comprises 41 samples. As before, the 400-metre-square area was divided into 50-metre-square blocks. However, this time the block values were estimated by the method of kriging. The range of influence of the semi-variogram model for this deposit was 100 m, so all samples within this distance of a block were included in its estimation. The results are shown in Fig. 5.5, and once again the upper figure in each block is the estimated value, whilst the lower is the kriging standard deviation, or standard error. For comparison, Fig. 5.6 shows the kriging solution for the case where the area is divided into 100-metre-square blocks. Notice that the kriging standard deviations in all cases are much lower than for the 50-

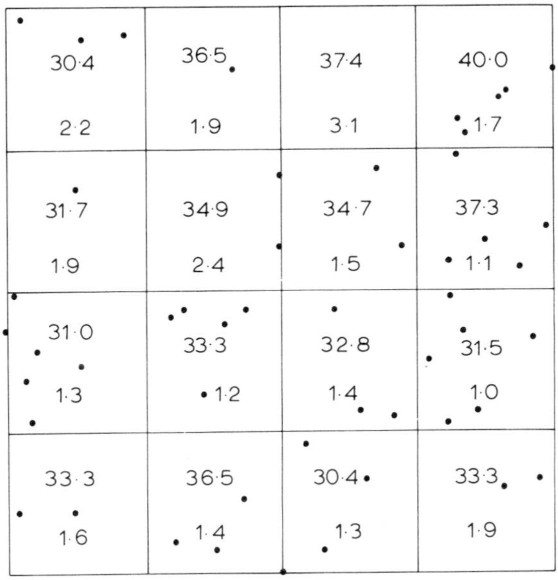

30·4 2·2	36·5 1·9	37·4 3·1	40·0 1·7
31·7 1·9	34·9 2·4	34·7 1·5	37·3 1·1
31·0 1·3	33·3 1·2	32·8 1·4	31·5 1·0
33·3 1·6	36·5 1·4	30·4 1·3	33·3 1·9

FIG. 5.6. Simulated iron ore deposit—block averages kriged from the set of random samples and the corresponding kriging standard deviations—100 metre blocks.

metre blocks. This once again bears out the principle that it is 'easier' to estimate large areas rather than small ones. Figure 5.7 shows the estimated values for the blocks when using the sample values from the regular grid. The kriging standard deviation for all of these blocks is 2·4 % Fe. This is not markedly different from the extension standard deviation of 2·5 % Fe. Perhaps our conclusion here must be that with a regular grid of this

31·9	31·0	36·5	38·3	34·1	34·5	38·1	37·0
30·8	28·1	33·1	33·4	29·8	34·2	38·2	35·8
33·6	30·6	31·7	32·2	32·3	36·5	36·8	34·0
33·9	32·5	33·6	36·6	36·9	36·4	35·0	34·3
33·3	32·4	33·8	36·2	36·3	35·2	32·7	32·3
35·2	36·0	36·9	34·2	33·4	36·0	34·6	32·8
35·0	35·1	37·5	34·6	31·0	33·6	36·3	34·9
37·5	32·0	34·2	37·0	33·7	31·6	33·9	34·1

FIG. 5.7. Simulated iron ore deposit—block averages kriged from the set of regular samples—50 metre blocks.

particular size, the arithmetic mean of the two 'corner' samples is as good an estimator as a weighted average of all samples within 100 m of the block. The exterior samples would seem to be superfluous in the circumstances. This conclusion does not hold for the irregular sampling, for which some great improvements in accuracy result from the application of kriging.

The usual requirement in ore reserve estimation is the production of block values. However, in many other possible applications of kriging—such as geochemistry or hydrology—the estimation desired is in the form of 'point' values, or a contour map of the variable of interest. Kriging can be used to produce the close grid of values necessary to the plotting of contour maps. In fact, the procedure is very much easier than 'area' estimation, since all the 'average semi-variogram values' reduce to simple values of the semi-variogram model itself. Since all of the observations are made at specified points, the left hand side of the kriging system is 'point-with-point' semi-variogram values. Since the value to be estimated is also at a point, the right hand side is also 'point-with-point' semi-variogram values. Figure 5.8 shows the contour map produced using the 50 randomly chosen samples. The blacked-out area at the top of the map is outside the range of influence

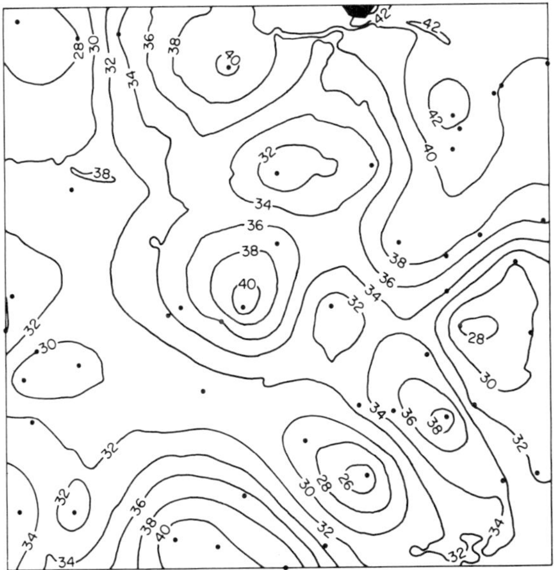

FIG. 5.8. Simulated iron ore deposit—contour map kriged from random samples.

FIG. 5.9. Simulated iron ore deposit—kriging standard deviation map for the kriged contour map from random samples.

of any sample, and hence cannot be estimated. One of the advantages of kriging as an interpolation technique is that every estimate is accompanied by a corresponding kriging standard deviation. Thus, for any contour map of values, a companion map of 'reliability' can be produced. This is shown in Fig. 5.9. The location of the sample points can easily be seen by the concentration of the low value contours in Fig. 5.9—the 1 % Fe and 2 % Fe contours. The highest possible contour value would be $5\sqrt{2} = 7{\cdot}07$ % Fe, which is the boundary around the blacked out area. This corresponds to

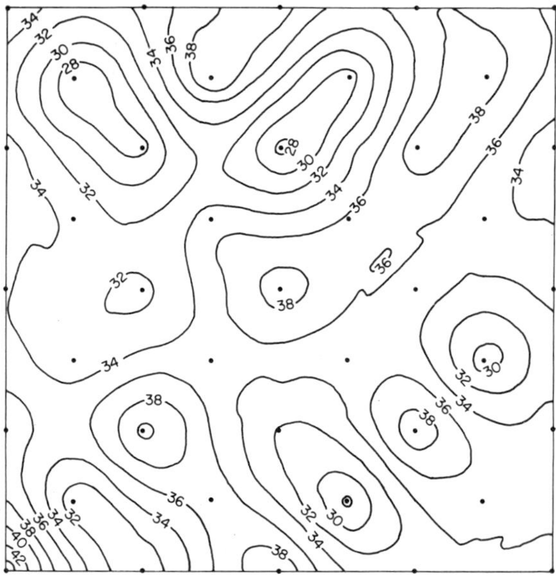

FIG. 5.10. Simulated iron ore deposit—contour map kriged from regular samples.

trying to estimate the value at a point which is just on the range of influence away from the nearest sample. The 'unreliable' areas are quite clearly outlined by the 5 % Fe contour (the sample standard deviation). A standard error which is larger than the original sample standard deviation denotes a rather unreliable prediction.

Figure 5.10 shows the interpolated contour map using the regular samples, and Fig. 5.11 the corresponding map of the kriging standard deviation. Note that, even though only 41 samples are available, and even

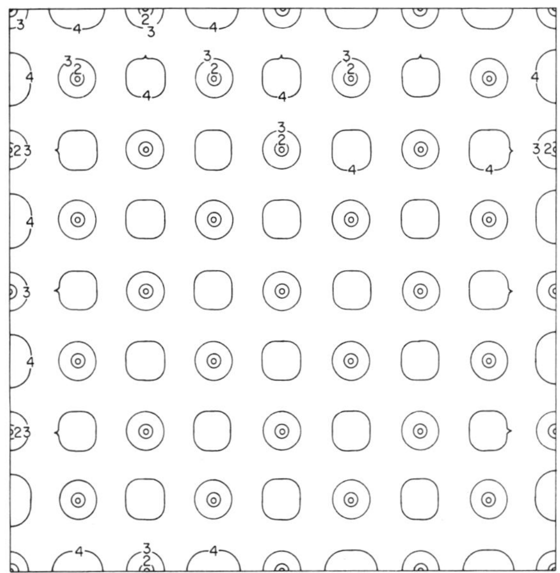

FIG. 5.11. Simulated iron ore deposit—map of kriging standard deviations for map kriged from regular samples.

though these are on a very large grid (71 m), the highest contour value in Fig. 5.11 is only 4 % Fe.

An additional advantage of kriging as an estimation technique is that the maps and/or calculations of the 'standard errors' can be produced *without* actually taking the samples. For example, if infill drilling were proposed on the regular grid, it is fairly obvious from Fig. 5.11 where the new samples should be taken. If a decision was taken to reduce the grid to 50 m, i.e. put a hole in the centre of each 4 % Fe contour, a complete new map of the resulting standard error could be drawn before setting foot in the field.

CHAPTER 6

Practice

Because this text was intended to be a 'first introduction' to the subject of Geostatistics, the examples and situations discussed have tended to be rather simplistic. There are many ore deposits—and other applications—which can be tackled with the methods described. However, there are at least as many others which cannot because of the presence of one or more complicating factors. In this chapter, I should like to mention *briefly* some of these problems, and perhaps indicate how they might be tackled. The order in which the problems are presented bears no relationship to their relative importance.

1. CONSTRUCTION OF SEMI-VARIOGRAMS USING IRREGULAR DATA

All the discussion on the construction of experimental semi-variograms in Chapter 2 was based on samples which were spaced more or less regularly within the deposit. Some grids had missing samples, but this presents no problems. In a situation where the samples are not regularly spaced, approximations must be introduced into the calculation. Suppose we wish to calculate the experimental semi-variogram value for a distance h in a specified direction (say, north-east). The chance of finding any pairs at exactly this separation with irregular sampling is quite small. We therefore place a 'tolerance' on each specification. We look for samples more-or-less distance h apart (within δh) and more-or-less north-east (within $\pm\delta\theta$)—see Fig. 6.1 for illustration. The size of the tolerances depends greatly on the structure of the deposit. This is rather a circular argument, since we do not know the structure until we construct the semi-variogram. If the deposit is anisotropic, the semi-variogram will be more sensitive to the tolerance

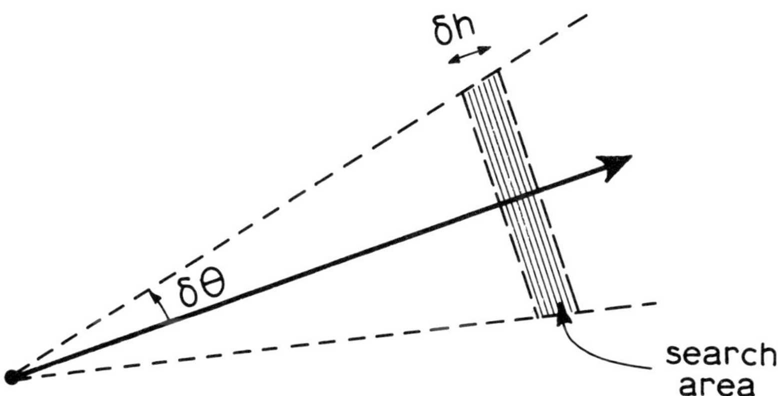

FIG. 6.1. Search area defined by tolerances on angle and distance between pair in experimental semi-variogram.

placed on the search angle. A good practice is to try several $\delta\theta$ values, and a narrow range of δh values. δh should always be small relative to the sample spacing. As a rule of thumb, you could try $\delta\theta = 5, 10, 20, 45$ degrees and δh equal to 10% of the average sample spacing.

2. SAMPLING ERRORS

This is a field which is skated over in most geostatistical treatises, and I intend to emulate my predecessors in this. *Random* errors introduced during sampling will contribute to the nugget effect in the semi-variogram, i.e. they will show as an increase in the 'unpredictable' component of the value. Similarly, core loss will also contribute to the nugget effect. Other possible contributions of 'error'—apart from the mineralisation itself—are analytical errors in valuation, subsampling and so on. However, the contribution of these errors should not be over-emphasised. In my Cornish tin example, we carried out a special sampling scheme to determine the 'size' of the operator error in the vanning assay. This turned out to be 3% of the total nugget effect. The other 97% was due to the random nature of the mineralisation.

Systematic errors, be they sampling, analytical or whatever, will *not* be picked up by Geostatistics and will be transferred to any estimates produced.

3. TRENDS

We have seen in Chapter 2 how to detect the presence of a significant trend within the deposit. If we construct the experimental semi-variogram assuming no trend, then the neglected component will show in the graph. If it is a periodic trend, it will show as a regular rise and fall in the semi-variogram. If it is a polynomial type of trend, it will show in the addition of a parabolic component to the 'true' semi-variogram. There are cases where a trend may exist but can be safely ignored, as in the silver example in Chapter 2. However, there are others in which this is not the case, e.g. the rainfall example. Kriging, as such, cannot be used in the presence of a strong trend. It will give erroneous and biased results. Some technique such as Universal Kriging, or the newer Generalised Covariances must be applied if the user is determined to use Geostatistics. My experience in the application of Geostatistics in the presence of trend is voluntarily non-existent.

4. ANISOTROPY

This is perhaps the easiest 'problem' to tackle. Most frequently, the form of the anisotropy is different ranges of influence in different directions. For example, a porphyry molybdenum may have a range of influence of 70 m vertically through the deposit, and one of 350 m in all horizontal directions. This is very simply tackled by changing the units of measurement in one direction so that the ranges *appear* to be equal. In the example cited, we might change all horizontal measurements to multiples of 5 m instead of 1. This gives a horizontal range of 70 units. Then, when estimating block values, it must be remembered that all horizontal distances must be expressed in units of 5 m. Diagonal distances must be corrected accordingly. For example, a block defined as 50 by 50 by 20 m, will be estimated as if it were 10 units by 10 units by 20 units. Similarly, the distance between samples must be corrected to this new 'co-ordinate' system. The simplest way to tackle this (inside a computer) is to correct all measurements before embarking on the estimation. The final estimates and standard errors will be as they should be, and need no 'uncorrection'.

5. IRREGULAR SHAPED STOPES OR BLOCKS

In the estimation problems discussed, all panels and blocks were rectangular in shape. The auxiliary functions are only relevant to such panels, and so cannot be used with, say, stope panels such as that shown in Fig. 6.2. These shapes can only be tackled using numerical approximations and a computer. The principle of the approximation is the same as that applied in calculating the three-dimensional F function in Chapter 3. Instead of considering all of the infinite number of points within the stope, we use a finite grid of points to represent it. The number of points is in

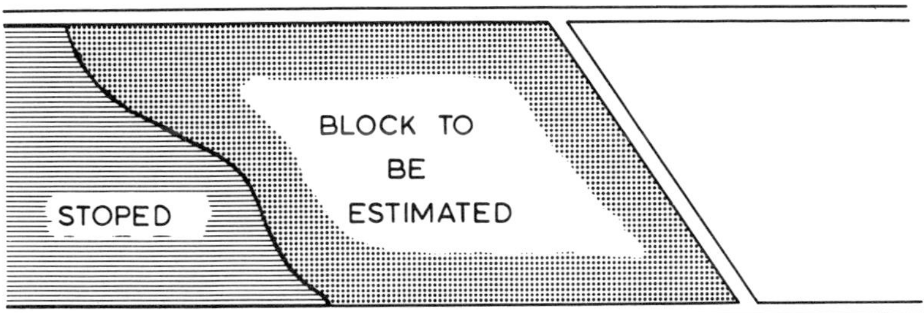

FIG. 6.2. Estimation of irregularly shaped stope.

question, but general agreement seems to lie in the range 64 to 100. This then means that values such as $\bar{\gamma}(S, A)$ are average semi-variograms between 'the sample and each point on the grid in the stope'. A computer program will evaluate the model semi-variogram value between each pair of points and then produce the average semi-variogram value required. This principle also applies to irregular shapes in three dimensions.

6. THREE-DIMENSIONAL KRIGING

This leads us on nicely to one of my hobby horses—the performance of kriging estimation in three dimensions. This problem is most often encountered in the planning of open pits from borehole results. The 'standard' technique is to make an approximation such as described in Section 5 above. A suggested alternative, which uses less computer time, is as follows:

(i) slice the deposit into benches;
(ii) represent each block as a panel at a level midway up the bench;
(iii) approximate this block by a grid of points in two dimensions;
(iv) take each borehole intersection with the bench and call this a 'point' midway up the bench;
(v) put all the slices back and krige.

The semi-variogram supposedly used in the kriging is that of 'bench composites', i.e. sections of length equal to the height of the bench. This technique is adequate if all boreholes are complete, and if they start and stop at more or less the same level. However, there are other situations which it does not represent adequately. Figure 6.3 illustrates some of these. All of these boreholes would be averaged over the bench, positioned

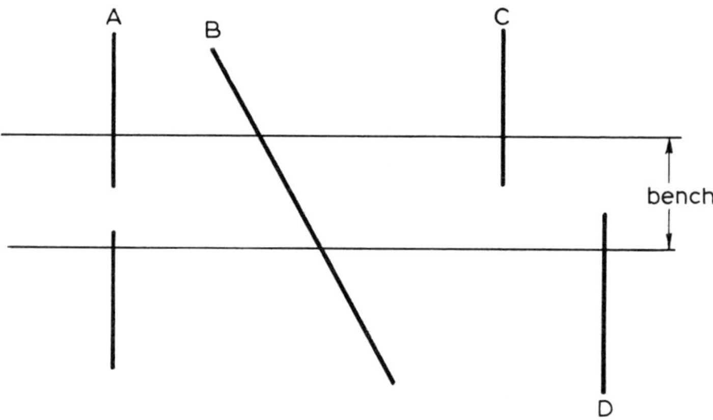

FIG. 6.3. Some problems with the simplistic approach to three-dimensional kriging procedures.

halfway up the bench and allocated a length equal to the height of the bench. Borehole A has core loss part way down the bench; borehole B is inclined so that the composite may be substantially longer than supposed; borehole C stops before it reaches the bottom of the bench and borehole D does not start until halfway through. All of these situations can be tackled by taking a truly three-dimensional approach to the problem. Computer packages are now available on the market for the above described method, the point approximation method, and the three-dimensional approach advocated in my own papers.

7. BIAS ON THE GRADE/TONNAGE CURVE

After a mine has been estimated on a block-by-block basis, it is usual to construct the so-called 'production' grade/tonnage curve. Unfortunately, even with the best estimates possible (kriging) this grade/tonnage curve will *still* be biased. There are two contributory factors to this bias. The first factor is that the selection criterion (cutoff value) is being applied to the *estimate* of the block grade. No matter how accurate that estimate, it will not exactly equal the *true* value of the block. Thus, if a block is estimated to be just below the cutoff value, there is a finite probability that the true value is above cutoff. However, this block will be sent as waste. On the other hand there will be blocks estimated as being above cutoff which are actually below cutoff. These will be sent as ore. Thus, we will have payable blocks sent to waste and unpayable ore sent to the mill. This will result in production figures which will differ from those predicted by the supposed grade/tonnage curve calculated on the block estimates. The major difference will be a lowering in the grade of ore milled.

The second bias in the grade/tonnage curve is one introduced by the volume–variance relationship. The estimates of the block values will not necessarily have the same variance as the actual block values. You may remember we discussed this matter in connection with the Cornish tin example in Chapter 3. There the estimator was the average grade of two 125-ft strips of ground, whilst the panel was 125 by 100 ft. The variance of these two quantities will not be the same. In most situations—except for point kriging—the variance of the estimator will be larger than that of the actual blocks. Thus the grade/tonnage curve based on the block estimates will be biased towards lower tonnage and an over-optimistic average grade. This problem is currently being investigated by the staff at Fountainebleau under the title 'Disjunctive Kriging'. A simple, empirically justified technique to correct this bias is currently under investigation at the Royal School of Mines.

SUMMARY

This summary is more for the book as a whole than for this chapter particularly. I have endeavoured to give a perhaps over-simplistic presentation of the theory and practice of the estimation technique known as kriging. Readers who find this approach tedious are referred to the definitive (and more mathematical) works mentioned in the bibliography. I have also endeavoured to detail the practical difficulties which arise in applying the technique and to suggest some ways of overcoming them.

Bibliography

The purpose of this bibliography is to detail some papers and books which might assist the 'new' reader to pursue the theory and applications of geostatistics. The references have been split by sub-headings to give a clearer notion of what the reader should hope to get from them.

1. Good introductory papers

These are fairly rare, and confined to the efforts of a few authors.

P. I. Brooker has published a few, such as:

'Robustness of geostatistical calculations: a case study', 1977. *Proc. Australasian Institution of Mining and Metallurgy*, vol. 264, pp. 61–8.

A. G. Royle is another good author, mostly publishing in the *Transactions of the Institution of Mining and Metallurgy*. One of his most recent papers is:

'Global estimates of ore reserves', 1977, vol. 86, pp. A9–17.

A. J. Sinclair has turned out two or three widely available papers, all concerned with explaining the application of geostatistics to particular deposits. Two such are:

'A geostatistical study of the Eagle copper vein, Northern British Columbia', *Canadian Inst. Min. Metall.*, 1974, vol. 67, no. 746, pp. 131–42.

'Geostatistical investigation of the Kutcho Creek deposit, Northern British Columbia', *Mathematical Geology*, 1978, vol. 10, no. 3, pp. 273–88.

2. Definitive volumes

Matheron, G. *The Theory of Regionalised Variables and its Applications*, 1971, Cahier No. 5, Centre de Morphologie Mathématique de Fontainebleau, 211 pp.

David, M. *Geostatistical Ore Reserve Estimation*, 1977, Elsevier, 364 pp.

Guarascio, M., David, M. and Huijbregts, C. (Eds) *Advanced Geostatistics in the Mining Industry*, 1976, D. Reidel, Dordrecht, Holland, 491 pp.

Journel, A. G. and Huijbregts, C. *Mining Geostatistics*, 1978, Academic Press, 600 pp.

3. Other introductory texts

Rendu, J-M. *An Introduction to Geostatistical Methods of Mineral Evaluation*, Monograph of the South African Inst. Min. Metall, 1978, 100 pp.

Royle, A. G. *A Practical Introduction to Geostatistics.* Course Notes of the University of Leeds, Dept. of Mining and Mineral Sciences, Leeds, 1971.

4. Other applications

Clark, M. W. and Thornes, J. B. *Forwards Estimation from Incomplete Data by the Theory of Regionalised Variables*, Non-Sequential Water Quality Records Project: Working Paper No. 4, 1975, LSE, 9 pp.

Delhomme, J. P. and Delfiner, P. *Application du krigeage a l'optimisation d'une campagne pluviometrique en zone aride*, Symposium on the Design of Water Resources Projects with Inadequate Data, 1973, UNESCO-WHO-IAHS, Madrid, Spain, pp. 191–210.

Huijbregts, C. *Courbes d'isovariance en cartographie automatique* Colloque sur la Visualisation, 1971, Nancy (Ecole National Superieure de Geologie), 11 pp.

Olea, R. A. 'Optimal contour mapping using universal kriging', *J. Geophys Res*, 1974, vol. 79, No. 5, pp. 696–702.

Poissonet, M., Millier, C. and Serra, J. *Morphologie mathematique et silviculture*, 1970, 3ieme Conference du groupe des Staticiens Forestiers, Paris. pp. 287–307.

5. Historical perspective

de Wijs, H. J. 'Method of successive differences applied to mine sampling', *Trans. Inst. Min. Metall.*, 1972, vol. 81, No. 788, pp. A129–32.

Journel, A. G. 'Geostatistics and sequential exploration', *Mining Engineering*, 1973, vol. 25, No. 10, pp. 44–8.

Krige, D. G. 'A statistical approach to some basic mine valuation problems on the Witwatersrand', *J. Chem. Metall and Min. Soc. South Africa*, 1951, vol. 52, No. 6, pp. 119–39.

Matheron, G. 'Principles of geostatistics', *Economic Geology*, 1963, vol. 58, pp. 1246–66.

Sichel, H. S. *The estimation of means and associated confidence limits for small samples from lognormal populations*, Symposium on mathematical statistics computer applications in ore valuation, S. Afr. Inst. Min. Metall, 1966, pp. 106–23.

6. Author's papers

These papers have been published to assist workers who wish to apply geostatistics to their own problems, and who wish to use a computer and/or numerical approximations.

'Some auxiliary functions for the spherical model of geostatistics', *Computers and Geosciences*, 1976, Vol. 1, No. 4, pp. 255–63.

'Some practical computational aspects of mine planning', in *Advanced Geostatistics in the Mining Industry*, ed. M. Guarascio *et al.*, 1976, pp. 391–9, D. Reidel, Dordrecht, Holland.

'Practical kriging in three dimensions', *Computers and Geosciences*, 1977, Vol. 3, No. 1, pp. 173–80.

'Regularisation of a semi-variogram', *Computers and Geosciences*, 1977, Vol. 3, No. 2, pp. 341–6.

7. Some useful journals to watch or read

The following Journals often contain papers on Geostatistics and related topics. Added to this list should be the Transactions or Proceedings of the numerous 'Institutions of Mining and Metallurgy', such as those of the UK, Canada, South Africa and Australasia.

Mathematical Geology
Computers and Geosciences
Economic Geology
Water Resources Research
Engineering and Mining Journal
Publications of the *Kansas Geological Survey*, especially the series on Spatial Analysis.
Proceedings of the annual *Applications of Computers in the Minerals Industry* (*APCOM*), various venues.
Canadian Inst. Min. Metall. Special Volumes.

Index